W0106057

Handbuch der mikroskopischen Anatomie des Menschen

Begründet von Wilhelm von Möllendorff

Fortgeführt von Wolfgang Bargmann

1. Band

Die lebendige Masse

3. Teil

Hans Georg Schwarzacher

Chromosomes

in Mitosis and Interphase

With 116 Figures

Springer-Verlag Berlin Heidelberg New York 1976

Professor Dr. Drs. h. c. Wolfgang Bargmann
Anatomisches Institut der Universität, 2300 Kiel, Neue Universität

Professor Dr. Hans Georg Schwarzacher
Histologisch-Embryologisches Institut der Universität, A-1090 Wien

Library of Congress Cataloging in Publication Data (Revised). Main entry under title: Handbuch der mikroskopischen Anatomie des Menschen, bearb. (   ) Begründet von Wilhelm v. Möllendorff: fortgeführt von Wolfgang Bargmann. Vol. 5, pt. 4 has title: Verdauungsapparat, Atmungsapparat. Vol. 6, pt. 6 has title: Blutgefäß- und Lymphgefäßapparat, innersekretorische Drüsen. Includes bibliographies. Contents: 1. Bd. Die lebendige Masse. T. 1 — 2. Bd. Die Gewebe. T. 1 — 3. Bd. Haut und Sinnesorgane. T. 1 — 4. Bd. Nervensystem. T. 1 — 5. Bd. Verdauungsapparat. T. 1 — 6. Bd. Blutgefäß- und Lymphgefäßapparat. T. 1 — 7. Bd. Harn- und Geschlechtsapparat. T. 1 — 1. Histology. I. Möllendorff, Wilhelm Hermann Wichard von, 1887 (ed.) II. Bargmann, Wolfgang, 1906 (ed.) [DNLM: 1. Histology. 2. Chromosomes— Ultrastructure. QS504 H236 Bd. 1]
QM551.H15   611 .018.   55-37658.
© by Springer-Verlag Berlin Heidelberg 1976
Softcover reprint of the hardcover 1st edition 1976
ISBN-13: 978-3-642-85912-0   e-ISBN-13: 978-3-642-85910-6
DOI: 10.1007/ 978-3-642-85910-6

Acknowledgements

My thanks are due to Dr. RENATE CZAKER and Doz. Dr. WOLFGANG SCHNEDL for valuable discussions and their help in many respects, especially in preparing some of the figures. The drawings stem from the skillful hand of Dr. WALTER HEIHS, to whom I am particularly grateful. Many colleagues were so kind as to give the permission to reproduce figures from their publications or to furnish me with original photographs. They are acknowledged individually in the text to the figures, as well as the publishers.

I am very grateful to Dr. PATRICIA FISCHER for her help in preparing the english manuscript. For technical assistance in making many new preparations for this monograph I want to thank Mrs. ANNA BRÓM, Mr. HANS-DIETER SCHERMANN, Mr. MIRZEA GURUIANU, and Mr. RUDOLF FIEDLER. Finally, I should like to express my sincerest thanks to Mrs. RICARDA WINTER for her excellent help in secretarial and bibliographical work.

The original investigations reported in this monograph for the first time were supported by grant No. 2514 of the "Fonds zur Förderung der wissenschaftlichen Forschung in Österreich".

Contents

I. Introduction

The progress in Micromorphology and Biochemistry of the last decades has led to a rather far reaching understanding of the function of the genes. Much is also known about their morphological organization within the cell, particularly their reduplication and segregation in connection with the process of cell division.

The intensive light microscopic studies of the earlier cytological era on cell division and chromosomes, which laid the basis for this understanding are very comprehensively covered by WASSERMANN (1929) in his masterly contribution "Wachstum und Vermehrung der lebendigen Masse" in this handbook.

There exist also many more recent reviews on chromosomes and on cytogenetics (e.g. SWANSON, 1960; MAZIA, 1961; TURPIN and LEJEUNE, 1965; WHITEHOUSE, 1969; HAMERTON, 1971; FORD, 1973). However, although some of them cover the more recent findings in man, they have either had to rely on more favorable species for detailed basic information or handled cytogenetic problems from a more practical and clinical point of view. Since moreover, the last few years have brought a flood of new information on chromosomes due to new cytological techniques, a new review on human chromosomes would seem justified within the frame of this handbook. This review will be restricted to *human somatic chromosomes,* i.e. it will leave out meiosis, and will provide information on other species only if this seems necessary for increased clarity.

It is hoped that this review will serve as a source of basic information not only for anatomists and cytologists, the traditional readers of this handbook but also for medical geneticists and clinicians.

II. Nomenclature and General Morphology of Chromosomes

Chromosomes are the dense and intensely stainable small bodies, first described by FLEMMING (1882), STRASBURGER (1882), and VAN BENEDEN (1883). The name "chromosome" was proposed by WALDEYER in 1888. Form and number of chromosomes are species specific. In general they appear as small threads or sticks about 1–2μ thick and several μ in length. In most species all cells of an individual have at least one complete set of chromosomes. In each mitotic division occurring in any tissue the same form and the same basic number of chromosomes are found. They are equally distributed to the daughter cells so that each contains the same chromosomes as the mother cell. Before the chromosomes divide they are replicated (or reduplicated) in the mother cell. The following description

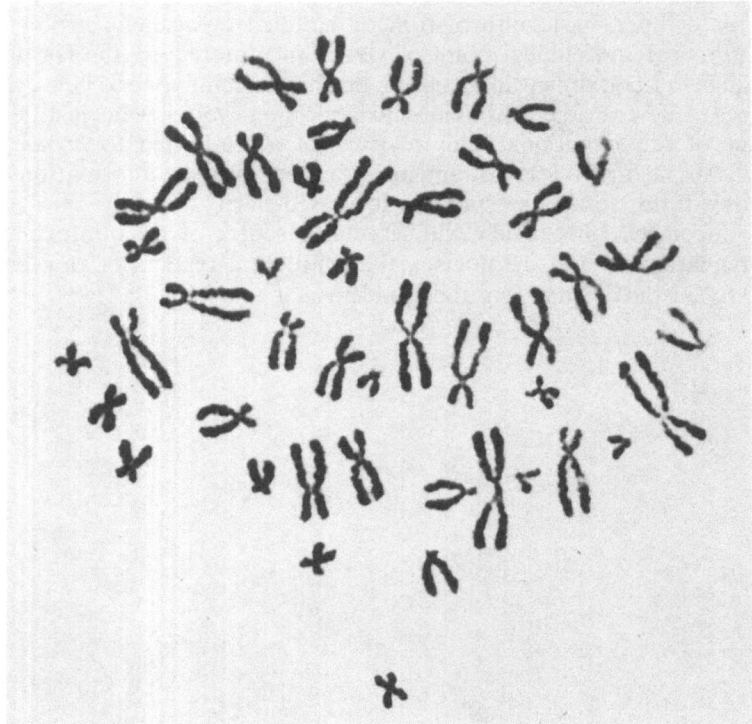

Fig. 1. Metaphase of a human cell (fibroblast culture) with spread chromosomes (Method similar to MOORHEAD *et al.*, 1960). Giemsa stain ×2000

of their general structure is based on the appearance in mitotic figures in standard cytogenetic preparations (Figs. 1 and 2).

Chromatids: After replication, each chromosome consists of two identical parts, called chromatids. During division these two chromatids are separated and one goes to each daughter cell. After division, a chromosome consists of only one chromatid and remains thus until again undergoing replication before the next mitosis. When entering mitosis a chromosome consists of two chromatids. Chromatids are therefore the microscopically visible units of replication.

Centromere and chromosome arms: After replication both chromatids are closely connected at one region, termed the centromere by DARLINGTON (1936). When the centromere divides at the onset of anaphase, the chromatids are finally separated. In the centromeric region also, the fibrils and microtubules of the mitotic spindle apparatus are connected with the chromosome by means of a special structure, the *kinetochore.*

In all chromosomes of man the centromeric region is rather small. In the light microscope it appears as a dotlike connection of the two sister chromatids. The chromatids are thinner at the centromeric region, which is therefore also referred to as the *primary constriction.*

The position of the centromere and the length of the chromosome are its most important distinguishing features. The centromere can be situated at any point between the end and the middle of the chromosome. Even before the term centromere was introduced, chromosomes were characterized by the position of their connection with the spindle fibres. WILSON (1928) defined two major types of chromosome 1. terminal or telomitic, and 2. nonterminal or atelomitic; nonterminal chromosomes may be median, submedian or subterminal. Currently the expressions telocentric, acrocentric, mediocentric or metacentric, submetacentric and subacrocentric are used (see LEVAN *et al.*, 1964).

The centromere divides the chromosome into two *arms.* In an exact mediocentric chromosome, both arms would be of equal length. A telocentric chromosome would consist of only one arm, all other types containing one long and one short arm. It is customary to designate the long arm with the letter "q" and the short arm with the letter "p". The position of the centromere can be defined by the ratio of the arm lengths. Three indices to express this are commonly used:

$$\text{Arm ratio, } r = \frac{\text{length of long arm}}{\text{length of short arm}};$$

$$\text{Centromere-index, } i = \frac{\text{length of short arm} \times 100}{\text{total length of chromosome}};$$

$$\text{Arm length difference, } d = \text{length of long arm} \atop \text{minus length of short arm}$$

The term "acrocentric" was introduced by WHITE (1945) to designate a chromosome which has the centromere "very close to one end". This term is preferable to "telocentric" because it has been questioned whether truly telocentric chromosomes can occur (RHOADES, 1940). None of the human chromosomes lack microscopically visible short arms.

Secondary constrictions: Some chromosomes show thinnings beside the centromere. Such areas are called secondary constrictions in contrast to the primary constrictions of the centromeric region. Length and degree of thinning is different and within certain limits characteristic for a given chromosome (Figs. 2 and 3). Some individual variability exists in this respect, and hence a variability between homologous chromosomes is possible (see Chapter IV). Some of the human chromosomes can, however, be quite clearly identified by their secondary constrictions (FERGUSON-SMITH *et al.*, 1962).

Secondary constrictions are closely related to heterochromatic regions and to nucleolus organizer regions. Their significance will be discussed in more detail in Chapter V.

Satellites: When a secondary constriction is situated near the end of a chromosome, the small distal piece of the chromosome may appear connected to the main piece only by a thin threadlike stalk (Fig. 2). Such a small piece is called a satellite. Five of the human chromosomes carry satellites. The size of the satellites is individually variable (see Chapter IV).

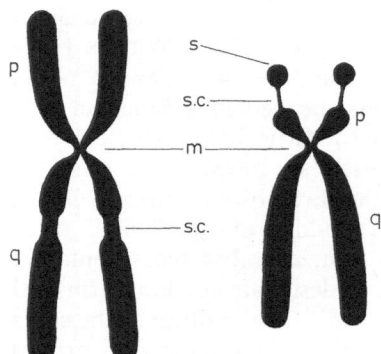

Fig. 2. Schematic drawing of a large metacentric and a satellite bearing acrocentric chromosome in metaphase. *m* centromere, *q* long arm, *p* short arm, *s.c.* secondary constriction, *s* satellite

Bands: With special methods of differential staining, bands can be demonstrated across the chromatids. The most important methods are: Staining with the fluorescent dyes quinacrine mustard or quinacrine dihydrochloride, which demonstrate the so called Q-bands (CASPERSSON *et al.*, 1970); Staining with Giemsa after special pretreatment (alkali, heat, proteolytic enzymes) demonstrating the so called G-bands (SUMNER *et al.*, 1971; SCHNEDL, 1971; DRETS and SHAW, 1971; DUTRILLAUX *et al.*, 1971). A few very distinct bands can be demonstrated by a treatment which supposedly de- and then renatures chromosomal

DNA. These are the so called C-bands (ARRIGHI and HSU, 1971). Details of these methods and their theoretical background are discussed in Chapters IV and V.

Chromomeres: This term has been used for the darker stained cross bands in polytene interphase chromosomes (the so called giant chromosomes of certain insects and plants). Chromomeres in this sense cannot be seen directly in mitotic chromosomes. Their relation to the chromosome bands produced by the special banding methods mentioned above will be discussed in Chapter V.

Chromonemata: A chromonema is a morphological unit indicating a fibril or a strand lying in the direction of length of the chromatid. So far, it cannot be decided with certainty whether a chromatid is built up of one or of more than one chromonema. In the first case which is by far the most probable the chromatid would be the only unit, that is single stranded, and in the latter case, it would be multistranded. The term "chromonema" is seldom used today and should preferably be abandoned.

Coils: The chromatids may show a helical coiling in later prophase, in metaphase, and sometimes even in anaphase (Figs. 1 and 70). The degree of coiling seems to be dependent on technical factors of preparation (OHNUKI, 1965). On the other hand it has been shown that the coils of human chromosomes are

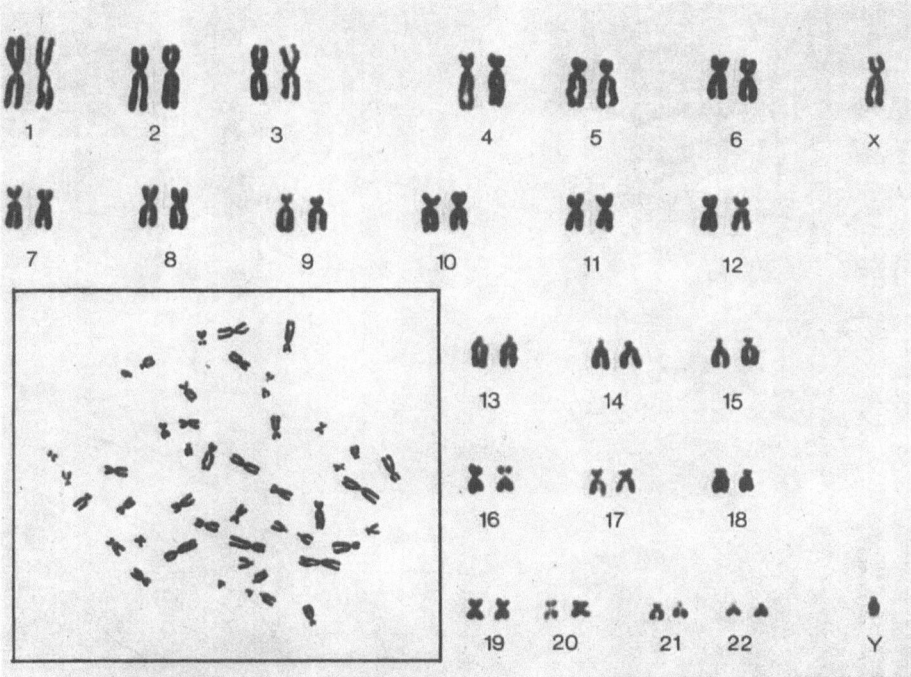

Fig. 3. Normal male karyotype. The chromosomes are arranged in pairs according to the Denver-London-system. Standard Giemsa stain. Individual identification is possible only for pairs *1–3* and *16–18,* and the Y chromosome. × 1500. Inset in the left corner: metaphase figure from which the karyotype was prepared. × 750. (From PASSARGE, 1974)

constant within certain limits and can even be used for the identification of single chromosomes (RUZICKA, 1973). The coils seen in the light microscope are sometimes referred to as "major coils". The assumption has been made that there exist in addition small coils, in fact that a chromatid is built up of a thin thread laid in a hierarchy of superimposed coils (see e.g. OSGOOD *et al.*, 1965). No morphological evidence has been found supporting this hypothesis (see Chapter VI).

Number of chromosomes: Man is a somatic diploid species with haploid gametes. The number of chromosomes per haploid set is 23 (TJIO and LEVAN, 1965; FORD and HAMERTON, 1956, see Figs. 3 and 4). Deviations from the haploid number of 23, except exact multiples of it, are called *aneuploidy*. If they occur in a greater proportion of the somatic cells of one individual, they are connected with certain clinical symptoms. They are known as numerical chromosome aberrations and are of considerable medical importance (see for review e.g. HAMERTON, 1971). In normal persons aneuploidy is found very rarely. COURT BROWN *et al.* (1966) reported an increase of cells showing loss of a single chromosome with age. In women over 60 years of age up to 7% of the blood cells may contain only 45 chromosomes, the missing one being presumably an X-chromosome. In men only up to 1–2% of the cells have one chromosome missing,

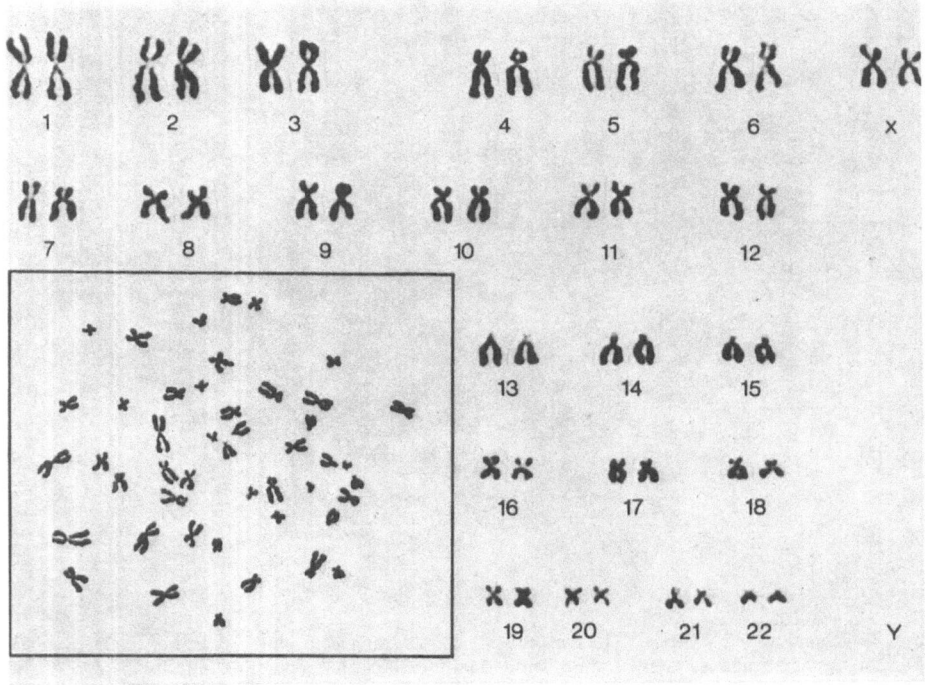

Fig. 4. Normal female karyotype, prepared as in Fig. 3 (From PASSARGE, 1974)

presumably the Y. Besides these changes no differences between individuals or between members of different races have been reported.

The term "diploid species" does not exclude the occurrence of *polyploid* somatic cells. In some tissues polyploid cells are regularly present, sometimes in high proportions. It should be remembered that e.g. in the liver, the occurrence of polyploid cells was long suspected on the basis of the observations on regular logarithmic classes of nuclear size (see e.g. JACOBJ, 1925). DNA-measurements and improved methods of chromosome counting supplied the evidence for the existence of polyploid cells (see e.g. HUSKINS, 1952; LEVAN and HAUSCHKA, 1953; VENDRELY and VENDRELY, 1956). In human cells *in vitro* the occurrence of poly-ploid cells is well known (FRACCARO *et al.*, 1960; SCHWARZACHER and SCHNEDL, 1965). Their origin by cell fusion or by endopolyploidization is discussed in detail by PERA (1970).

Normally only multiples of the diploid chromosome number are found in somatic cells, but the sporadic occurrence of triploid cells in blood cultures derived from normal persons has been reported by PAWLOWITZKI and CENANI, 1967). Other investigators, however, point out that triploid or haploid cells are never found in normal tissues (e.g. EDWARDS *et al.*, 1967). In tissue cultures from other mammals the occasional occurrence of haploid, triploid and hexaploid cells has been reported (PERA and SCHWARZACHER, 1969; PERA and RAINER, 1973; RIZZONI *et al.*, 1974). Also a predominantly triploid cell line from *Microtus agrestis* has been described (PERA and SCHOLZ, 1974).

III. Chromosome Morphology during Mitotic Phases

1. Introduction

In the previous chapter the general morphologic features of chromosomes were described, based on their appearance in mitosis, specially in metaphase. In this chapter changes taking place during the course of mitosis will be considered.

2. The Mitotic Phases

The course of mitosis is in principal similar in most types of cells. Countless observations on human tissues have been made during the years, and most of them are still valid today. Attention should be drawn to the excellent older as well as more modern reviews (e.g. WASSERMANN, 1929; GEITLER, 1938; SWANSON, 1960; MAZIA, 1961; FORD, 1973).

The mitotic phases are, shortly described, the following:

The *prophase* is characterized by a shortening and thickening of chromosomes. They become visible as thin threads which are irregularly folded. In late prophase, the double structure of the chromosomes, i.e. the two chromatids are usually visible.

In the *metaphase*, the chromosomes are shortened and thickened so much that they appear as small sticks. The two chromatids are clearly seen, also the centromeres, the secondary constrictions, the satellites, and other characteristic features. In metaphase, the chromosomes assume a characteristic position in one plane in the center of the cell, forming the so called "metaphase plate". The nuclear membrane has disintegrated shortly before and the spindle apparatus was built up. Two poles are formed, the metaphase plate of the chromosomes lying in the middle between them. The nucleoli have usually disappeared by the beginning of metaphase.

In the *anaphase* the two chromatids of each chromosome segregate and the two daughter groups of chromosomes (each consisting of only one chromatid) go to the two opposite poles. The chromosomes are even thicker and shorter than in metaphase.

In the *telophase* the thick and swollen looking chromosomes form two confluent groups near the poles. A nuclear membrane is formed around each group. The cytoplasm divides, and in the *reconstruction phase* the transition to two normally appearing daughter cells is completed.

The duration of the mitotic phases is in the vicinity of one hour. It is mainly known from *in vitro* studies. Culture medium and temperature are of considerable influence. Differences may also exist *in vivo*.

3. Changes of Chromosome Morphology during Mitosis

The method of preparation is of great influence on the appearance of chromosomes. Direct observations of living cells with the phase contrast microscope gives, of course, a true picture of chromosome structure, but the chromosomes are frequently covered by other cell components or by themselves (Fig. 5). In preparations selectively stained for chromosomes (as e.g. by the Feulgen reaction or with haematoxylin, Figs. 6–9) their form can be more easily studied. The process of shortening during the course of mitosis is clearly visible. More details of chromosome structure are shown, if preparations are pretreated with a hypotonic solution, fixed in acetic acid-alcohol and then air dried or squashed. This method introduced by Hsu (1952) and Hughes (1952) is today widely used as a routine procedure for cytogenetic studies (for detailed descriptions and discussions of different variants of the method see Yunis, 1965, and Schwarzacher and Wolf, 1974). In Figs. 10–17 a series of mitotic stages prepared by this method, is shown.

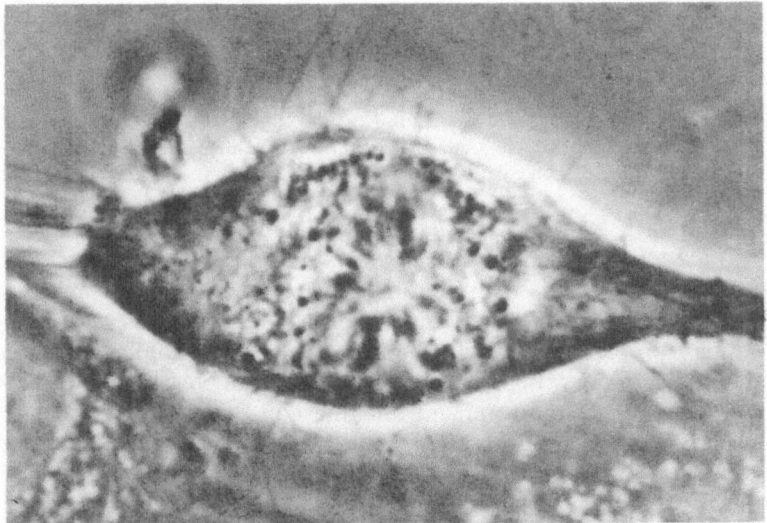

Fig. 5. Living human cell from a fibroblast culture in metaphase. Phase contrast. × 2000

In early prophase, (Fig. 10) the chromosomes are still rather long and thin. A complete spreading of all chromosomes is not possible although usually some of them may lie isolated and hence can be studied. The position of the centromeres can be already recognized in favorable preparations. An identification of some of the chromosomes is therefore possible. As will be shown in Chapter IV, identification of all the chromosomes is possible in prophase by the banding methods. The two chromatids are visible only in favourable cases. In later stages of prophase (Fig. 11) the chromosomes are shorter, thicker and rather straight. The two chromatids become clearly visible but are still frequently twisted together.

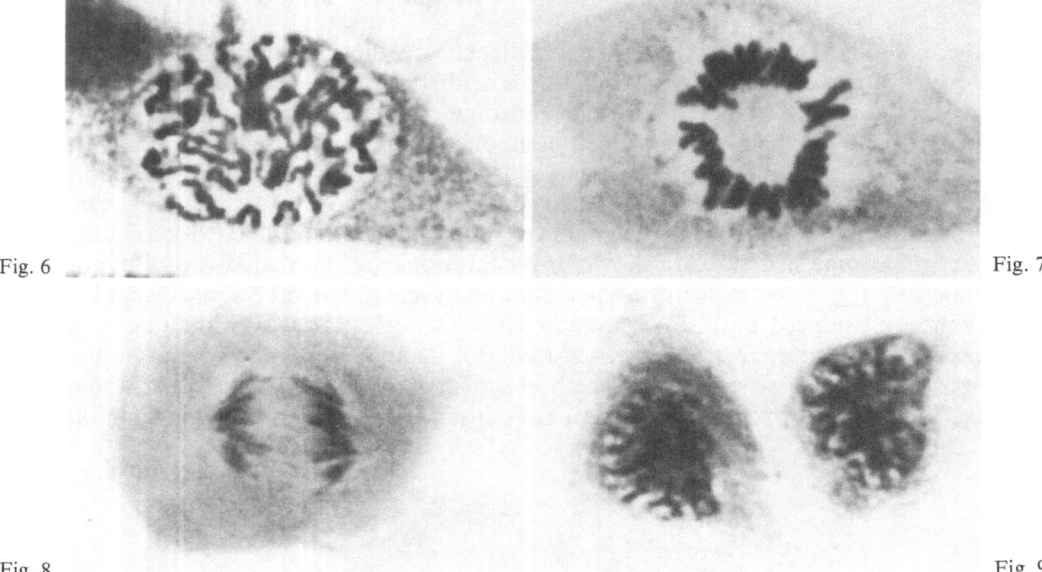

Fig. 6   Fig. 7

Fig. 8   Fig. 9

Figs. 6–9. Human cells from a fibroblast culture in the four main stades of mitosis: Prophase
(6), metaphase (7), anaphase (8), telo-reconstructionphase (9). Haematoxylin-Eosin stain, fixation
in ethanol-acetic acid (3:1). × 1200

Primary and secondary constrictions can be seen as well as satellites. According
to DARLINGTON (1936) the contraction of the chromosomes is achieved by a
"spiralization" process which may involve even several successive orders of coiling. In human preparations, it can be seen that early prophase chromosomes
show only a light coiling with a relatively small coil diameter. Towards metaphase,
the coiling becomes more prominent and the coil diameter increases (Figs. 12
and 13). Recent electron microscope studies have revealed that the number of
coils decreases in the same relation as the diameter of the coils increases during
the spiralization cycle (RUZICKA, 1973).

In prophase of female cells, one of the two X-chromosomes is sometimes
more contracted and thicker than all the other chromosomes (Fig. 11). Frequently
it appears to be also more tightly folded. This X-chromosome is the heterocyclic
and inactivated one (see Chapter VII). Its behaviour resembles precocious condensation, the classical sign of heterochromatin (HEITZ, 1929).

Towards the metaphase, the chromosomes look like sticks or short stretched
threads (Figs. 12 and 13). In addition to their characteristic morphological features
the two sister chromatids are very clearly visible. The heterocyclic X-chromosome
is sometimes recognisable because of its particularly parallel chromatids. The
same is true for the Y-chromosome (SCHMID, 1967).

The segregation of the two chromatids becomes more and more distinct.
They give the impression of being forced apart but are still connected at the
centromere (Fig. 13).

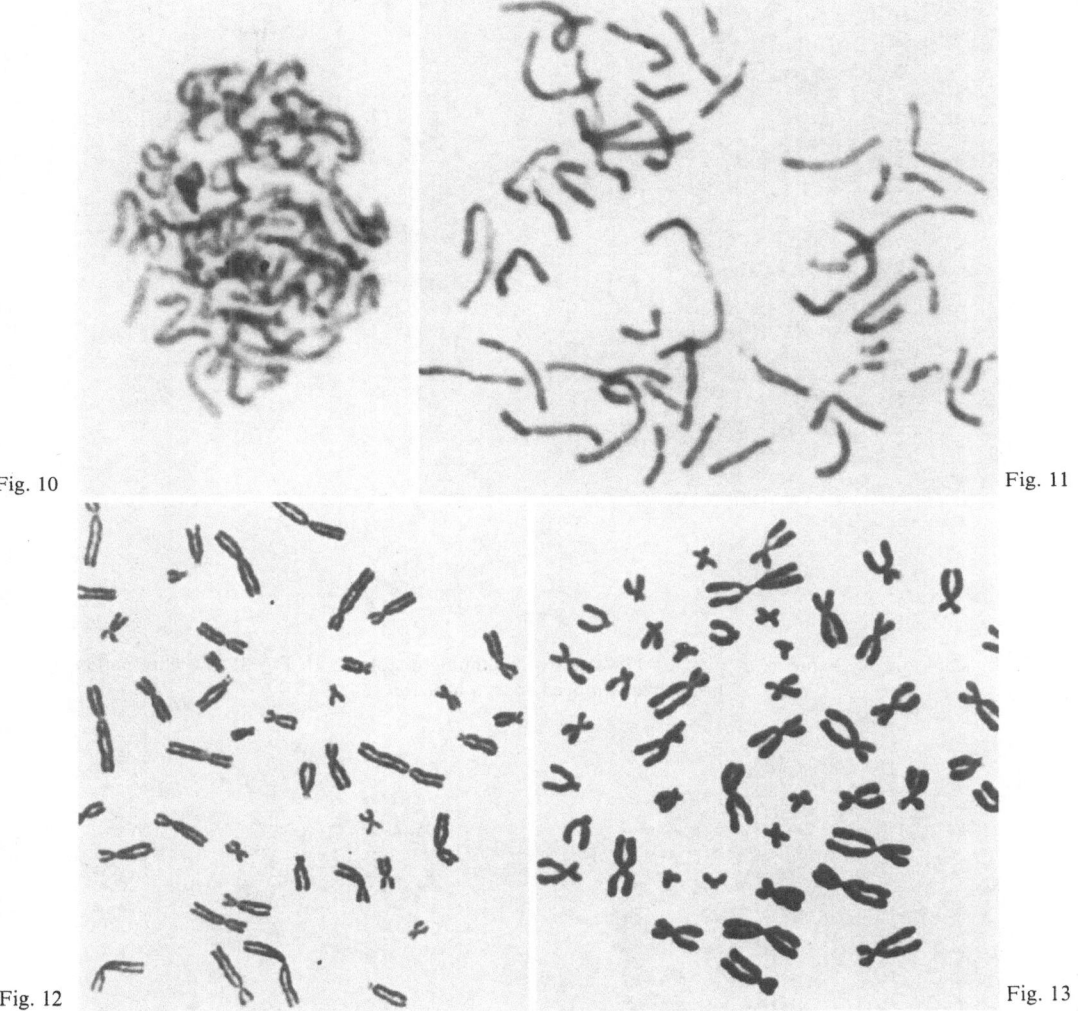

Fig. 10

Fig. 11

Fig. 12

Fig. 13

Figs. 10–13. Human cells from a blood culture in different stades of mitosis: Early prophase (10), late prophase (11), early metaphase (12), late metaphase (13). Pretreatment with hypotonic solution, fixation in ethanol-acetic acid, air dried. × 1200

The beginning of anaphase is defined by the moving apart of the chromatids. The connection of the two chromatids at the centromere is obviously dissolved quite suddenly. In early anaphase the chromatids are still lying closely together (Fig. 14). They contract still further during their way towards the poles (Fig. 15).

When the two daughter groups of chromosomes are formed, telophase begins. The chromosomes are now even thicker and shorter than in anaphase. With advancing telophase they look like swollen knobs, becoming less dense and more and more diffuse in outline (Figs. 16 and 17).

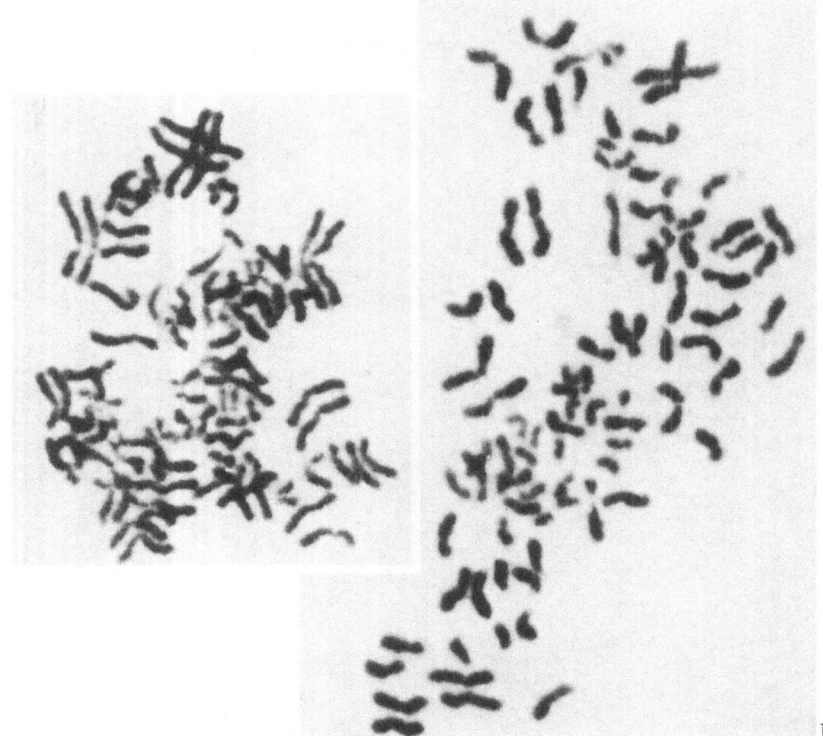

Fig. 14

Fig. 15

Figs. 14 and 15. Same as in Figs. 10–13, stades of beginning anaphase. In Fig. 15 the chromatids have already moved apart a little farther

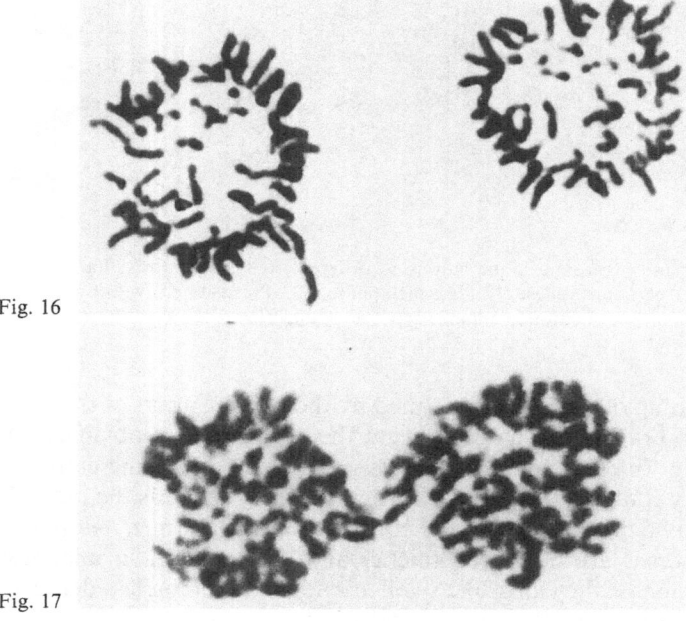

Fig. 16

Fig. 17

Figs. 16 and 17. Same as in Figs. 10–15, stades of telophase

It is apparent that the decondensation of the chromosomes in the telo- and the reconstruction phases is not just the reverse process of the condensation during prophase. During prophase the chromosomes start as very thin and long threads rather distinctly outlined, whereas in telophase they end as diffuse lumps.

4. Differential Contractions of Chromosomes

During prophase not all the chromosomes may shorten to the same degree. The example of the heterochromatic X-chromosome in female cells has already been mentioned. It is of course difficult to measure the shortening of the single chromosomes within a given absolute period of time. This is only possible by direct observations of mitosis in living tissues. Human cells are not suited for this. Particularly favourable for such studies are endosperm cells of *Haemanthus* (BAJER, 1959) where the individual chromosomes can be observed and measured during the course of mitosis. In Fig. 18 the measurements by BAJER (1959) are given. As can be seen, the speed of contraction is not constant during prophase and metaphase. The different chromosomes contract, however, at approximately

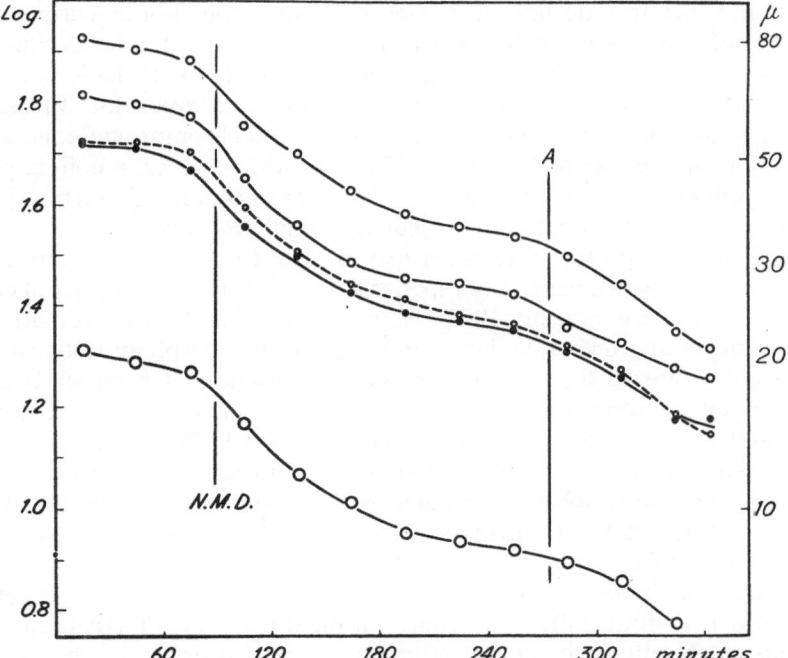

Fig. 18. Chromosome length (plotted on a logarithmic scale) of living *Haemanthus*-cells. Each of the four upper curves represents mean values of three homologous chromosomes of the larger chromosomes. The lowest curve mean values of the six smallest chromosomes. *N.M.D.* nuclear membrane disappearance. *A* beginning of anaphase. (From BAJER, 1959)

the same rate. As BAJER (1959) points out, the pattern of coiling and therefore the rate of shortening seems to differ between different plant species, and this is probably also true for animal species.

Reports on length measurements of chromosomes in different mitotic phases from fixed preparations are somewhat contradictory. WICKBOM (1949) found in different amphibians and fishes that long chromosomes contract relatively more than short chromosomes toward the end of prophase and during metaphase, but that at early prophase this pattern is reversed. In this way, a compensation occurs and the chromosomes shorten fairly proportionally to their length over the whole process of contraction. In human mitosis SASAKI (1961), FITZGERALD (1965), SMITH (1965), and GERHARDT (1970) found a relatively greater contraction of long chromosomes than of the short chromosomes. LEVAN et al. (1962) reported a uniform contraction of long and short chromosomes in mouse cells. The length differences of mouse chromosomes are, however, not as great as those of human chromosomes, and measurements may be not so exact. MATTHEY (1962) also noticed no difference in contraction between long and short chromosomes. These conflicting results may be explained by the fact that in human routine cytogenetic preparations only a short period of the contraction process can be observed. In reality all chromosomes may contract proportionally to their length, but at different points of time to a different degree, as indicated by WICKBOM (1949). Another possibility would be that the longer chromosomes undergo an additional coiling, and therefore do indeed shorten more than the shorter chromosomes. This point of view was originally taken by DARLINGTON (1933) and is also discussed by FITZGERALD (1965). The observation of PATAU (1960) that the long chromosomes are thicker in late metaphase than the shorter chromosomes further supports these authors. Our present knowledge of the chromosome coils as seen in the electron microscope (RUZICKA, 1973) favour, however, a coiling process which is rather constant for long and short chromosomes. It seems that the problem of differential contraction is not yet completely solved.

Some of the reported differences may be due to the different amounts of heterochromatin which varies from species to species for chromosomes of comparable lengths (see e.g. SCHMID, 1967). HEITZ (1929) called heterochromatin chromosomal material, which is similarly condensed during interphase and metaphase. Later studies showed that this interphasic condensation of heterochromatin is in general not as strong as the metaphasic contraction of chromosomes. However, BOYES and SLATIS (1954) and BOYES and NAYLOR (1962) showed very clearly that in insects the heterochromatic parts of the chromosomes are indeed much more condensed in prophase than euchromatic chromosomes. Similar observations were made in the Chinese hamster by HSU and ZENZES (1964) and by COREY et al. (1967)

In man, the facultative heterochromatic X-chromosome in female cells has been shown to contract differently from euchromatin. As has already been pointed out, this X chromosome can sometimes be seen in prophase as being clearly more darkly stained and more folded than the other chromosomes. Many investigators have reported that the length of the allocyclic X in late prophase and in metaphase is variable (e.g. GILBERT et al., 1961; BADER et al., 1963; ATKINS et al., 1963; GERMAN, 1964; COURT BROWN et al., 1964). However, PATAU (1965) and BISHOP *et al.* (1965) found, that if only metaphases are considered, these variations are not greater than for other chromosomes. In an extensive study GERHARDT (1970) presented evidence that on the average, the late labeling X-chromosome (that is the facultative heterochromatic X) is relatively shorter in early prophase and relatively longer in late metaphase than the other chromosomes of the C-group. This precocious condensation is however small, and is moreover not found in all cells. This indicates differences in the allocyclic condensation of the inactive X-chromosome from cell to cell which may also be reflected in the wellknown occurrence of female interphase cells lacking an X-chromatin body (see Chapter VII).

RÖHME (1974) reported that in chromosomes of the Indian muntjak the regions which contain few G-bands condense more than regions containing relatively many G-bands. RÖHME (1974) investigated not only prophase and metaphase chromosomes but also premature condensed chromosomes (experimentally achieved by fusion of a cell in metaphase with one in interphase, see Chapter VII). The so called "G-bands" are presumably also regions of a relatively high density (see Chapter V). Therefore the same principle as in the case of heterochromatin would be valid: relatively dense chromosome regions condense relatively little.

IV. The Human Karyotype

1. Introduction

The representation of the systematically arranged chromosomes of a cell is called the karyotype of the cell (Figs. 3 and 4). In a broader sense the term karyotype is also used in reference to a particular individual inferring that all the (somatic) cells have the same karyotype. Moreover the term can be used for a species, if individual variations which are very small compared to species differences are excluded. In this sense the expression "human karyotype" is used here.

The arrangement of the chromosomes is made according to their size and form. Certain rules of nomenclature, and for arranging the chromosomes of a human cell have been laid down in international conventions (Denver Report, 1960; London Report, 1963; Chicago Conference, 1966; Paris Conference, 1971). In this chapter the normal human karyotype will first be described, as it can be constructed by means of conventional or standard cytogenetic preparations. This will be followed by a description of the karyotype as revealed by the refined methods for demonstrating chromosome banding. Finally the individual variabilities will be considered.

2. Simple Staining Methods

It has already been mentioned that cytogenetic standard preparations are made by means of a hypotonic pretreatment of the cells followed by a fixation in acetic acid-alcohol and by a spreading of the cells either by air drying or by squashing. The simple or conventional staining methods make use of acetic orcein, Giemsa staining solution, or Fuchsin (for detailed information on chromosome techniques see YUNIS, 1965; and SCHWARZACHER and WOLF, 1974). With these staining methods the outline and the most characteristic structures of chromosomes can be seen (Figs. 1, 3 and 4). Late prophase and metaphase are particularly suitable for constructing karyotypes. This is usually done from photomicrographs or camera lucida drawings, from which the chromosomes have been cut out and arranged. There exist many reviews describing the standard karyotype (e.g. PATAU, 1965; HAMERTON, 1971; PASSARGE, 1974). Therefore only a very brief description is necessary here.

The main criteria for arranging the chromosomes are the length and the position of the centromere. The autosome pairs are numbered from 1 to 22 in order of size as accurately as possible. The sex chromosomes are designated X and Y. As will be discussed in the next paragraph, special staining methods allow the identification of each chromosome. Currently the numbering given by CASPERSSON *et al.* (1970, 1971) with the aid of the quinacrine fluorescence staining is generally accepted, although it does not strictly follow the order of chromosome size.

According to their form (centromere position) the chromosomes can be divided into 7 groups, designated by the letters A to G. In Figs. 3 and 4 each chromosome pair is labelled with a certain number. For most chromosomes, this is arbitrary since only 6 pairs and the Y can be definitely identified by the simple staining methods. In detail, the chromosomes can be described as follows:

Group A: The 3 largest chromosomes form this group. They all can be individually identified. No. 1 is almost exactly mediocentric (centromere index about 48–49). In the longer arm a secondary constriction near the centromere is sometimes visible. No. 2 is submetacentric, with a centromere index of 37–40. No. 3 is again nearly metacentric, the centromeric index being about 45–47.

Group B: This group comprises the chromosome pairs 4 and 5. Both are submetacentric (centromere index 24–30). They can usually be easily differentiated from the chromosomes of other groups, but not from each other.

Group C: In this group chromosomes 6–12 and the X-chromosome are included. They are all submetacentric, the centromeric index ranging from 28 to 40. With conventional simple methods it is sometimes possible to identify No. 6 (relatively metacentric and distinctly larger than the rest), and No. 9 (carrying a secondary constriction on the long arm near the centromere).

Group D: Three acrocentric chromosome pairs, Nos. 13–15, are included in this group. The centromeric index is about 15. They cannot be differentiated from each other. All carry satellites but usually not all of them are seen in a particular cell.

Group E: This comprises chromosomes Nos. 16–18. No. 16 is metacentric (centromere index about 40). No. 17 is subacrocentric (centromere index on the average about 25), and No. 18 slightly less subacrocentric. All three can be differentiated from each other.

Group F: This comprises 2 pairs of small metacentric chromosomes, No. 19 and 20. Their centromeric index is around 40. They cannot be differentiated from each other with simple staining methods.

Group G: Two small acrocentric chromosomes, Nos. 21 and 22, are included here with a centromeric index of about 20–25. Both carry satellites but again these cannot usually be seen in one cell.

Y-chromosome: According to its size and centromere position the Y-chromosome belongs to group G, but it carries no satellites. Its long arm shows a considerable individual variation.

3. Special Staining Methods

It has been shown in recent years that structural differences exist along the length of chromosomes which can be made visible by special staining methods. These differences appear as cross bands and are therefore referred to as "bands" and the methods bringing them out as "banding techniques". Three groups of such techniques are most important for the identification of single chromosomes.

C-bands

The Paris-Conference (1971) designated as "C-staining methods" those which demonstrate "constitutive heterochromatin", and as a "C-band" a unit of chromatin stained by these methods.

The term "heterochromatin" is used to describe certain parts of the chromosomes which behave in a different manner from all other chromatin, the latter being called "euchromatin". "Constitutive" heterochromatin is a type which is rather invariably found on both homologues of an autosome and on the Y chromosome. The (secondary) inactivated X-chromosome of a female cell is an example of another type, called "facultative" heterochromatin (see Chapter VIII).

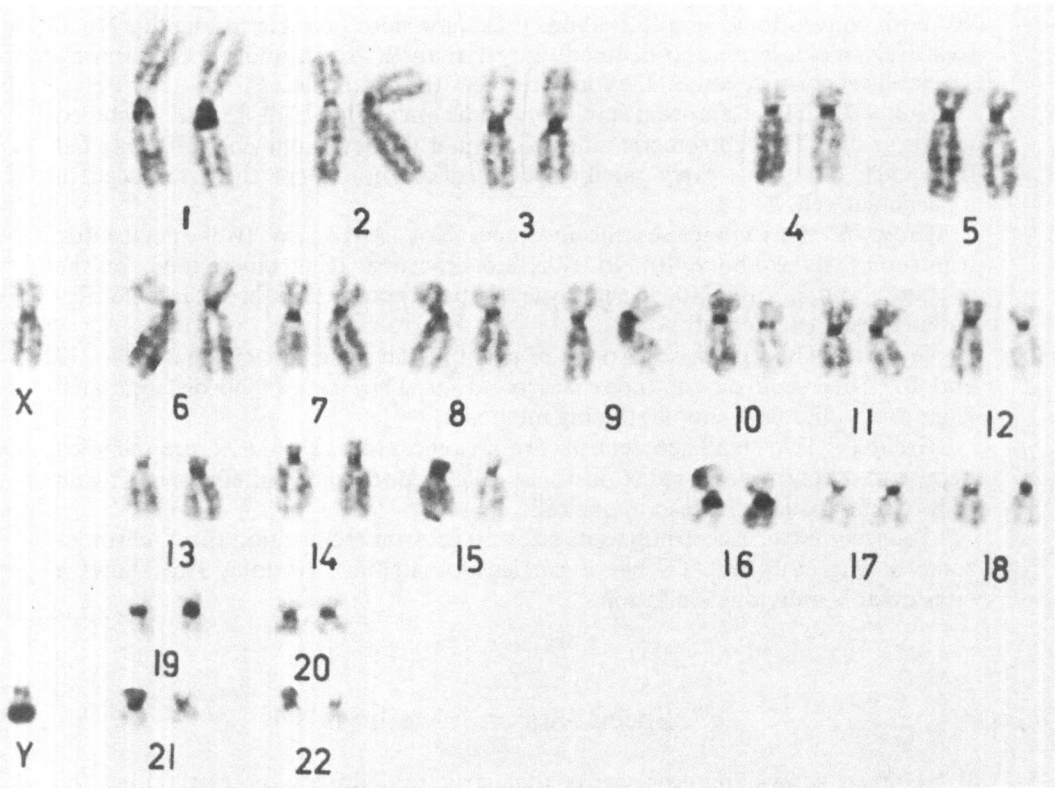

Fig. 19. Karyotype from a male cell. Staining after the methode of ARRIGHI and HSU (1971) for constitutive heterochromatin. Ca. × 2000

The C-staining methods were developed from the procedure used to localize satellite DNA containing highly repetitive sequences (PARDUE and GALL, 1969, 1970; JONES, 1970). The first description of human chromosomes stained with such a method was given by ARRIGHI and HSU (1971) and YUNIS *et al.* (1971). The technique is thought to depend on the denaturation of DNA in fixed chromosome preparations by treatment with NaOH or heat, followed by a renaturation in warm saline solution. The repetitious DNA in the C-band regions may renature more rapidly than the rest of the DNA and therefore stain deeply with Giemsa. The mechanism of C-band staining is discussed in more detail in Chapter V. Details on the technical procedure can be found in SCHNEDL (1974).

In Fig. 19 a karyotype stained by the C-banding method is presented. It can be seen that the centromeric regions of the chromosomes are intensely stained. The size of the bands near the centromere is not the same in all chromosomes. Only very small spots are seen for instance, in chromosomes No. 2 and Y.

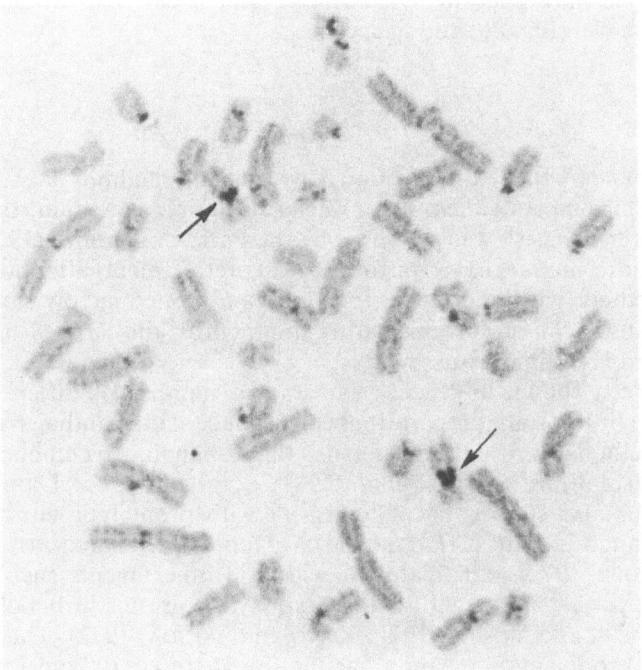

Fig. 20. Metaphase of a female cell, stained after the method of BOBROW *et al.* (1972). Large deeply stained blockes on both chromosomes No. 9. Ca. × 1 500

In some chromosomes other regions besides the centromeric bands are also deeply stained. These other regions correspond to the secondary constrictions seen in simple stained standard preparations (compare with Fig. 3 and 4). Particularly to note are chromosomes 1, 9, and 16. The distal part of the long arm of the Y-chromosome also shows dense staining. This part as well as the secondary constrictions show individual variations in size.

The exact position of the centromeric C-bands in regard to the centromere cannot be given with certainty in each chromosome. They lie adjacent to the centromere in most cases.

With a special technique some of the heterochromatic C-bands can be made particularly distinctly visible. If a standard chromosome preparation is stained with a Giemsa solution at pH 11.0–11.6 for a certain time, these bands appear red against a pale blue of the rest of the chromosomes (BOBROW et al., 1972; GAGNÈ and LABERGE, 1972). This method is called "Giemsa-11-method". The most prominent band stained is the secondary constriction of chromosome No. 9. If proper staining times are used, this band will be shown almost selectively (Fig. 20). With prolonged staining times other C-band regions as for instance those of chromosomes No. 1, 5, 7, 13–15, 17, 20, 21, and 22 are stained. In addition the distal part of the Y-chromosome appears sometimes reddish. The C-band of chromosome No. 9 is so large that it can also be seen in somatic interphase nuclei (Fig. 86) and in sperms.

G-bands

The finer bands which can be shown by various methods are designated by this term. Since most of them use a Giemsa-solution as a stain, the letter "G" standing for "Giemsa" was proposed by the Paris-Conference (1971). Although naming these techniques (which afford special pretreatment) after quite a common staining method (widely used e.g. for blood and bone marrow smears and also in the C-band technique) seems rather unpractical, the term G-band has been accepted and is commonly used.

It has been shown in recent years that a number of different treatments of standard chromosome preparations can produce a fine banding pattern. Several of these methods were developed from the C-banding techniques, using heat or alkaline solutions (SUMNER et al., 1971; SCHNEDL, 1971; DRETS and SHAW, 1971). It was also shown that a treatment with proteolytic enzymes produces the same banding pattern (DUTRILLAUX et al., 1971; SEABRIGHT, 1972; WANG and FEDOROFF, 1972) and treatment with still other agents such as e.g. urea, potassium permanganate, and detergents may also result in banding (UTAKOJI, 1972; SHIRAISHI and YOSIDA, 1972; KATO and YOSIDA, 1972). The first reported method for producing fine bands was that by DUTRILLAUX and LEJEUNE (1971). It stains just those regions which stay unstained by all other methods and vice versa. The bands thus produced are therefore also referred to as "reverse" or "R-bands".

The cause and the theoretical background for the staining of fine bands is discussed in the next chapter. In the following only a brief description of the G-bands (the fine bandings) is given.

The G-bands are present on all chromosomes along the whole length of their arms. Many of the G-band techniques do not stain the regions of the C-bands, but those using alkaline solutions as pretreatment show them as well.

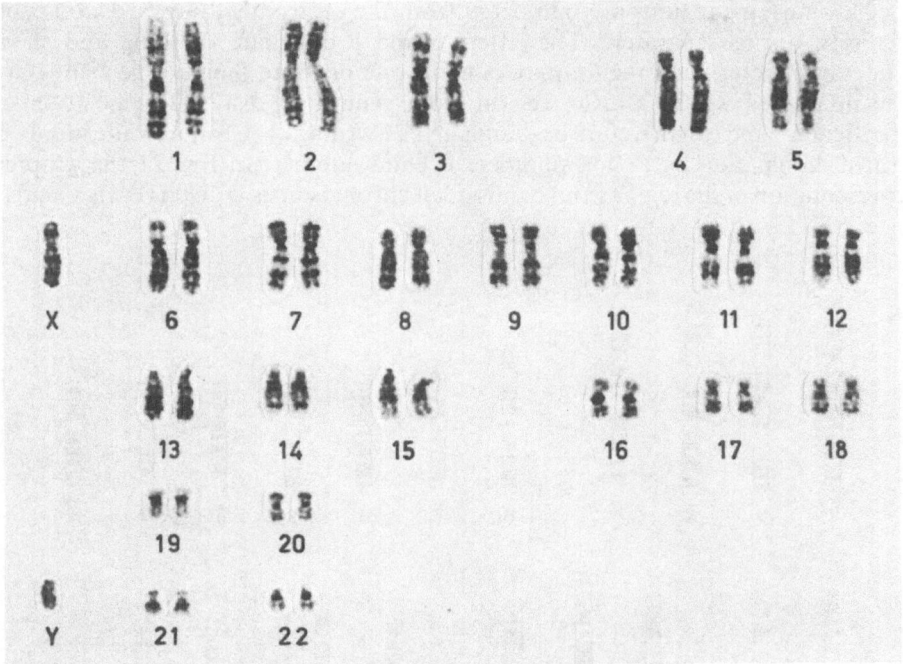

Fig. 21. Karyotype from a male cell. G-band staining after the method of SCHNEDL (1971).
Ca. × 1 500. Courtesy of W. SCHNEDL

In Fig. 21 a karyotype is given of a metaphase stained by the method of SCHNEDL (1971) which is of the latter type. It can be seen that it is possible to identify each chromosome beyond doubt because of their very characteristic banding patterns. The chromosomes are numbered according to CASPERSSON (1970). Instead of a description of all the bands a graphic representation, as published by SCHNEDL (1971) is shown in Fig. 22.

Terminology of the bands

The Paris Conference (1971) and its Standing Committee on Chromosome Band Nomenclature proposed a system of terminology of the bands: "A *band* is defined as a part of the chromosome which is clearly distinguishable from its adjacent segments." Any given area of a chromosome has to belong to a band, and no interband regions exist.

A *landmark* is defined as any distinct morphological feature that can be used in identifying a chromosome. Characteristic bands, centromeres, secondary constrictions and satellites are considered to be such landmarks. Variable features are not taken as landmarks. A chromosome *region* is defined as the area between the mid-lines of two adjacent landmarks.

Chromosome regions are numbered from the centromere outwards, on each of the two arms separately. The letters q and p designate the long and short arm respectively. Each region may contain one or more bands. The bands are also numbered within a given region from central to distal. The position of a particular band can therefore be designated by writing: 1. Chromosome number, 2. arm designation, 3. region number, 4. band number. In Fig. 23 the graphic representation of the regions and bands of all chromosomes is given (Paris Conference, 1971).

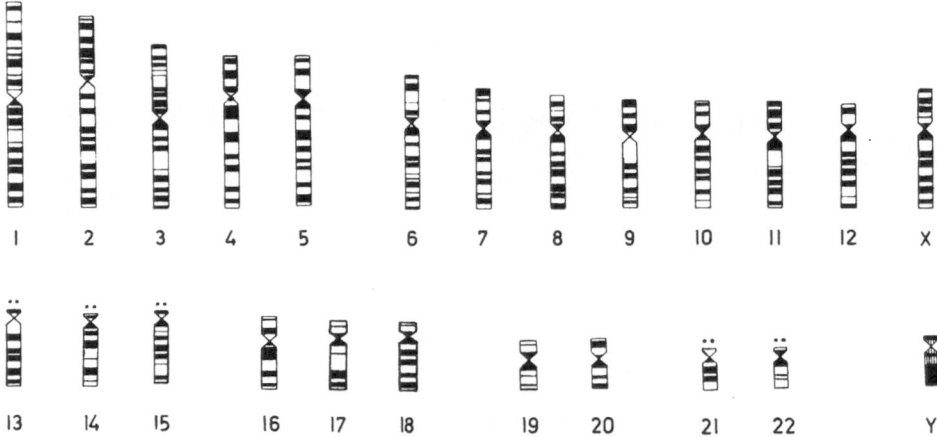

Fig. 22. Schematic representation of the G-bands in human chromosomes. Courtesy of W. SCHNEDL

A comparison between Fig. 22 (SCHNEDL) and Fig. 23 (Committee) shows that in the latter fewer bands are considered. This is partly because the Committee on Band Nomenclature has tried to designate only the very distinct bands hitherto known, seen by all methods including the so called Q-banding. On the other hand the map prepared by SCHNEDL (1971) includes consistent fine bands which are seen in prophase chromosomes. Fig. 24 shows such a prophase and the great number of bands is indeed impressive. About 250 bands can be visualized microscopically (SCHNEDL, 1971) and up to 500 by electron microscopy (BAHR, 1973) in the haploid set. These bands fuse during chromosome contraction thus forming the coarser bands of metaphase chromosomes.

The Committee on Band Nomenclature proposes to use additional numbers for an eventually necessary subdivision of a band. It is also noteworthy that the methods which do not include an NaOH-treatment or which use proteolytic enzymes usually cannot produce so many fine bands (Fig. 25). These methods are, however, simpler to perform and therefore of a very high practical value. The very fine bands are not always necessary for identification of the chromosomes (e.g. SPERLING and WIESNER, 1972; SUN et al., 1973).

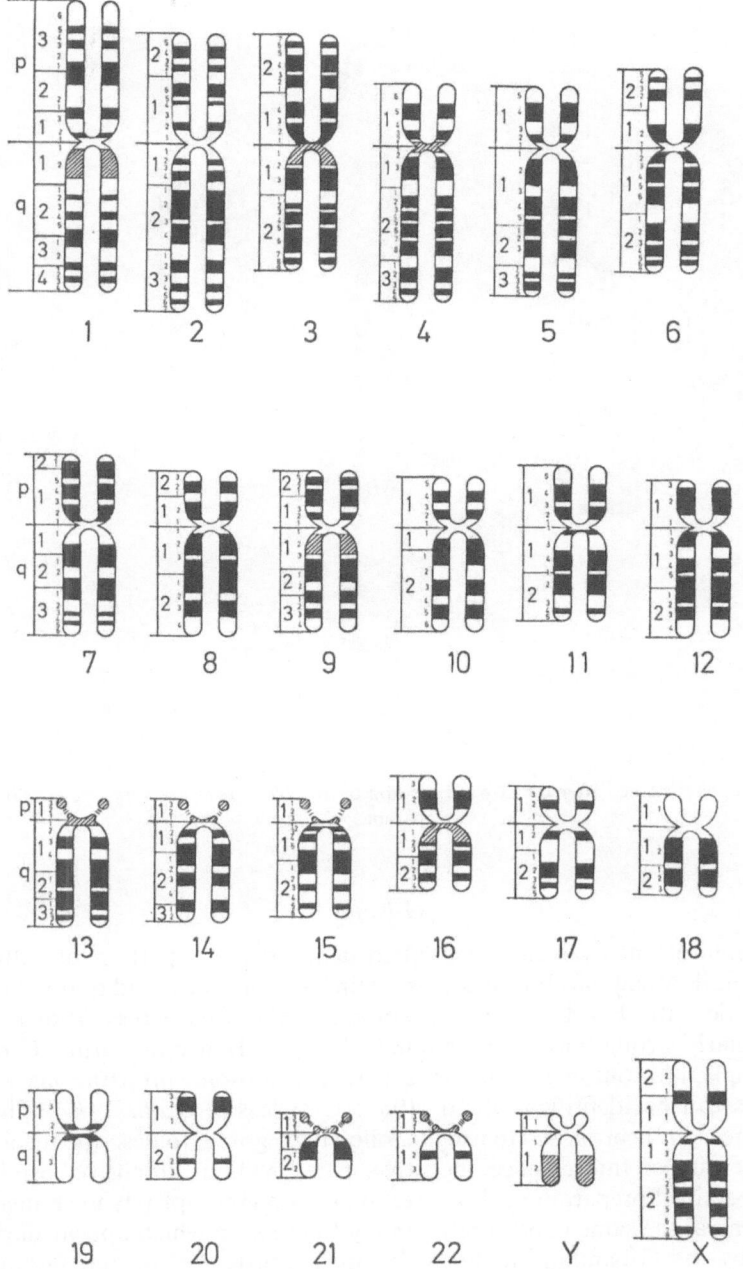

Fig. 23. Schematic representation according to the Paris nomenclature of the bands and regions in human chromosomes as observed with the Q-, G-, or R-staining methods. Positive Q and G-bands, and negative R-bands are shown in black, variable regions are shown hatched. Region numbers (large) and band numbers (small) as provided by the Paris Conference. From PASSARGE (1974) after the Paris Conference (1971). Courtesy of The National Foundation, New York

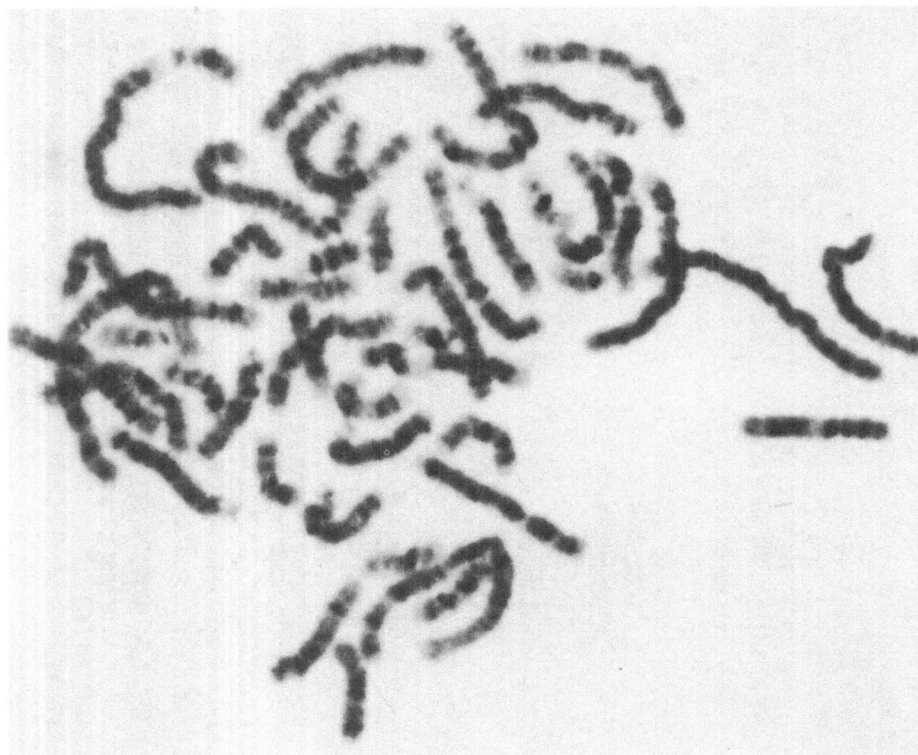

Fig. 24. Prophase stained after the G-band method of SCHNEDL (1971). A very fine banding pattern
is apparent. × 2200

Q-bands

Several fluorescent dyes can be used to demonstrate a pattern of differential
stained bands along the chromosomes. Quinacrine mustard and quinacrine dihydrochloride, introduced by CASPERSSON et al. (1968) have been shown to give
a particularly strong fluorescent staining (Fig. 26). Hence the term "Q-bands".
When applied to routine human cytogenetic chromosome preparations, all chromosomes can be identified (ZECH, 1969; CASPERSSON et al., 1970). The Paris
Conference (1971) proposed to use the following 5 grades to describe the approxi-
mate intensity of fluorescence: negative, pale, medium, intensive, brilliant. In
properly stained preparations, however, no region is completely unstained.

In general the same bands are intensely fluorescing which appear dark when
stained by the G-banding methods. In other words the Q- and G-bands are
almost identical with a few exceptions. The most prominent exceptions are the
negative Q-staining of the secondary constrictions of chromosomes 1 and 16,
and the particularly brilliant Q-staining of the Y-chromosome (distal part of
the long arm). Some regions show a marked individual variability. This polymorphism will be discussed in the following paragraph.

The brilliantly fluorescing part of the Y-chromosome is so large that it can be seen in interphase nuclei (PEARSON et al., 1970; CASPERSSON et al., 1970). It appears as a small very intensely fluorescing spot (Fig. 83), called the "Y-chromatin" (in contrast to the "X-chromatin" or the so called "Barr-body", see Chapter VII). Normally no other region shows a fluorescence of comparable brilliancy and of comparable size. Fluorescence staining with quinacrine of interphase nuclei is of great practical value since it allows the diagnosis of the presence of a Y-chromosome.

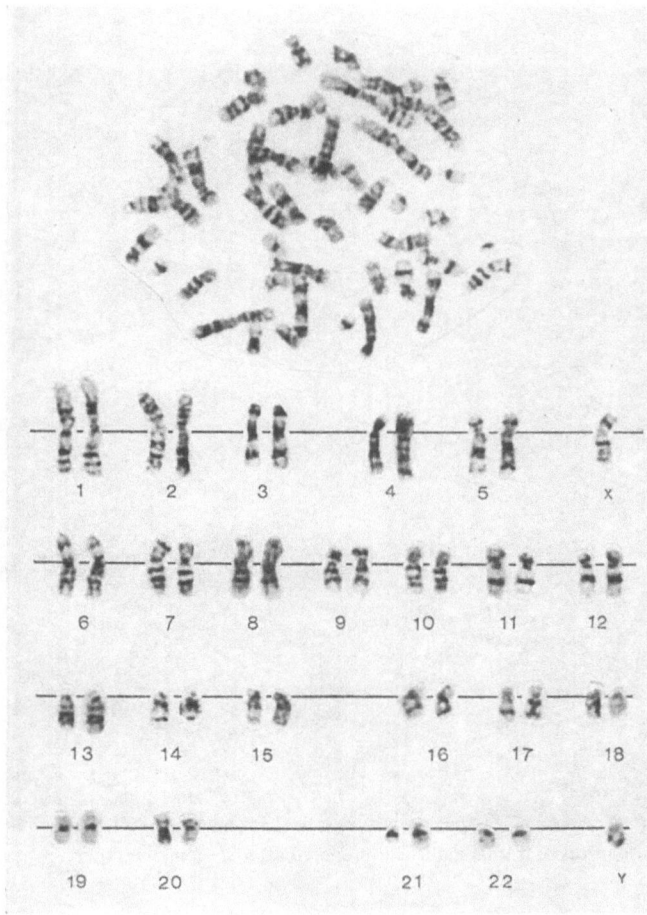

Fig. 25. Metaphase and karyotype of a male cell after pretreatment with trypsin followed by Giemsa stain. (From PASSARGE, 1974)

R-Bands

If standard chromosome preparations are incubated at 87° C in a low concentrated salt solution and then stained with Giemsa, the same regions which are weakly stained by the G-banding methods appear darkly stained (DUTRILLAUX and

LEJEUNE, 1971). Since the chromosomes reveal a pattern reverse to the G-banding pattern, these bands are called "R-bands". The R-banding method can of course be of the same importance in identifying chromosomes as the other banding methods.

A somewhat prolonged pretreatment of the preparations can lead to a particularly strong staining both with Giemsa and Quinacrine of the ends of the chromatids (DUTRILLAUX, 1973). This has been referred to as "T-bands" (telomeric bands). T-bands can be of practical value in certain cases of translocations where the ends of chromosomes are involved (DUTRILLAUX, 1973).

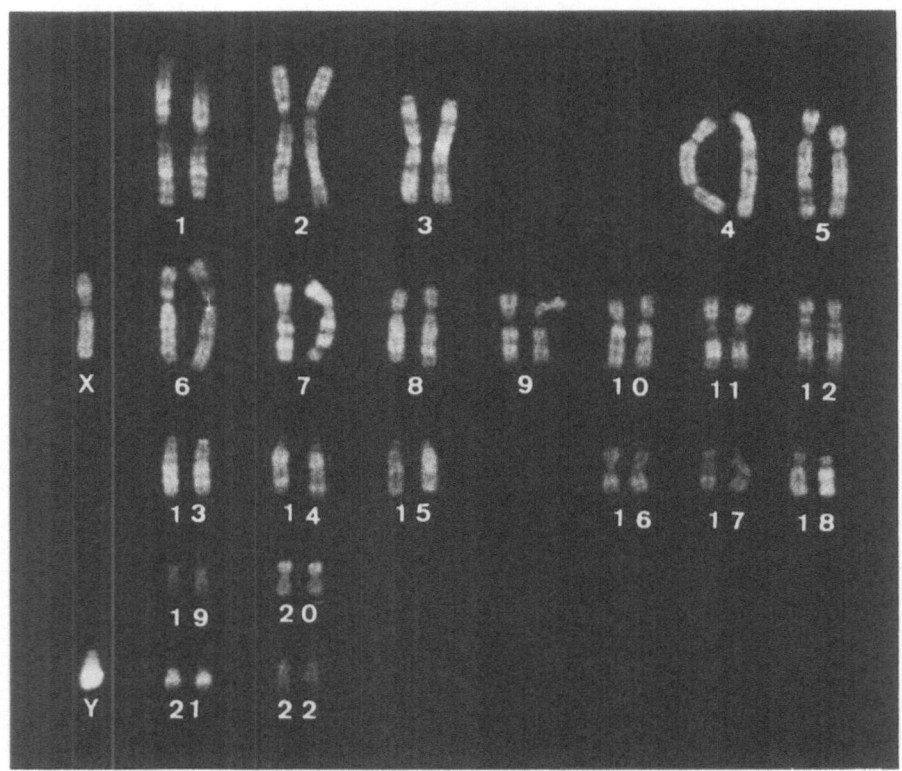

Fig. 26. Karyotype from a male cell, prepared after staining with quinacrine mustard

4. Individual Variations

Individual variations in *chromosome number* have already been discussed in Chapter II.

Variations in *Chromosome structure* in phenotypically normal individuals exist too. So far as is known up to now, these variations are the same in all cells of an individual. No tissue specific or function specific variation can be detected with the morphological techniques available today. It is clear that since such variations have by definition no, or only small effects on the phenotype, they must concern chromosomal areas which are genetically inactive or of rather low activity.

Chromosome breaks are observed in a small percentage in preparations of phenotypically normal persons. LITTLEFIELD and GOH (1973) found lesions of chromosomes in 5.6% of male cells and in 6.5% of female cells in a total of 29705 metaphases from 122 blood cultures from men and 183 cultures from women. The sex difference is statistically significant. Other factors influencing the frequency of chromosome breaks are obviously the age of the cultures and different laboratory conditions, such as temperature, composition of the culture medium, light and radiation.

A number of *structural rearrangements* of chromosomes in phenotypically normal persons are also known. They are mainly translocations (see e.g. COURT BROWN, 1967). Although they may not lead to any phenotypical abnormality of somatic cells, such variations must be considered as pathological, since they may lead to abnormal meiosis and to chromosomally abnormal gametes.

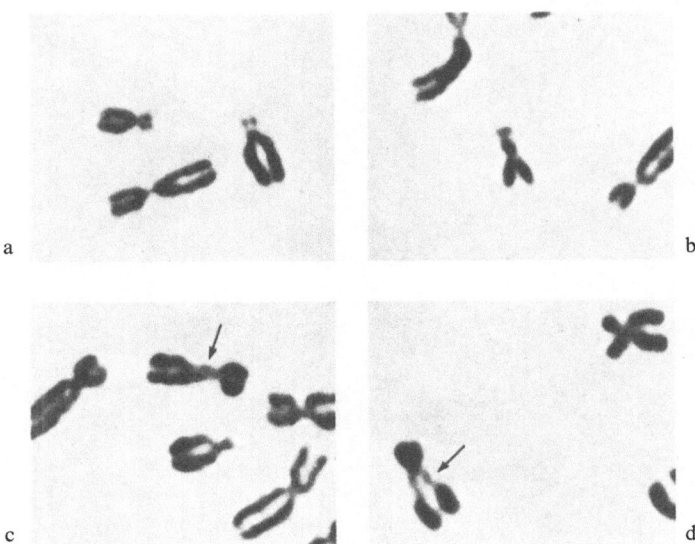

Fig. 27a–d. Variants of chromosomes of the D group (a), G group (b), and prominent secondary constriction of a chromosome No. 9 (c and d). Standard Giemsa stain. (From PASSARGE, 1974)

Secondary Constrictions

Even with simple staining methods of routine chromosome preparations an individual variability of the size of secondary constrictions can be observed. This
has been thoroughly studied in several publications (e.g. FERGUSON-SMITH *et al.*,
1962; TURPIN and LEJEUNE, 1965; LUBS and RUDDLE, 1971)

The most prominent and variable constrictions are in chromosomes 1, 9
(MADAN and BOBROW, 1974) and 16 (Fig. 27). Variabilities can sometimes be
seen on other secondary constrictions, but these are more clearly observed with
the aid of the banding techniques, particularly Q-banding. A few areas, corresponding to secondary constrictions (for instance in chromosome No. 3), show
a very distinct variability in quinacrine fluorescence (see below). The secondary
constrictions on chromosome No. 9 (band q12) can be studied particularly well
by the "Giemsa 11"-technique (Fig. 20).

Acrocentric Chromosomes and Satellites

Any of the ten acrocentric chromosomes (13, 14, 15, 21, 22) may bear satellites
(FERGUSON-SMITH and HANDMAKER, 1963). It is, however individually variable
how many of the acrocentric chromosomes bear them as seen in standard preparations, and how big they are. Particularly prominent variabilities of satellites
have been followed through two or several generations of a family (COOPER
and HIRSCHORN, 1962; ELLIS and PENROSE, 1965; ENGMANN, 1971). They have
been reported in identical twins and in the members of a highly inbred group
of people (BAHR and GOLOMB, 1971). There is no doubt that they are inherited
in a Mendelian fashion.

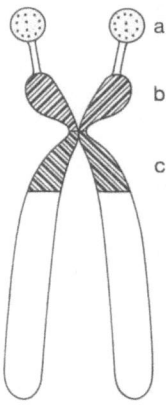

Fig. 28. Schematic representation of the three possible sites of variants *a, b, c* in acrocentric chromosomes. (After SCHNEDL, 1974b)

The fine banding methods, particularly quinacrine fluorescence staining, allow a more detailed description of the variabilities and an exact attribution to certain chromosomes (MIKELSAAR et al., 1973; SCHNEDL, 1974; see Fig. 29).

Variations on an acrocentric chromosome can occur on any of the four following areas (see Fig. 28). a) the centromeric region of the long arm; b) the proximal part of the short arm; c) the secondary constriction of the short arm (which represents the "satellite stalk"); d) the satellite (SCHNEDL, 1974). These four regions correspond to the following regions and bands of the Paris Nomenclature (compare with Fig. 23): a) q11 plus central part of p11; b) proximal part of p11; c) p12; d) p13. In accordance with the recommendation of the Nomenclature Committee of the Paris Conference (1972) one should subdivide p11 into a part belonging to the centromeric region, and a part belonging to the short arm outside the centromeric region. One should therefore write: a) q11 plus p11,1; b) p11,2.

The variations of these areas concern size as well as fluorescence intensity after quinacrine staining (MÜLLER and KLINGER, 1974). For instance the satellites may be quite large, or they may be so small as to be invisible in microscopic preparations, or they may perhaps even not be present at all. The intensity of the fluorescence may also vary (Fig. 29). Of course any valuation is subjective.

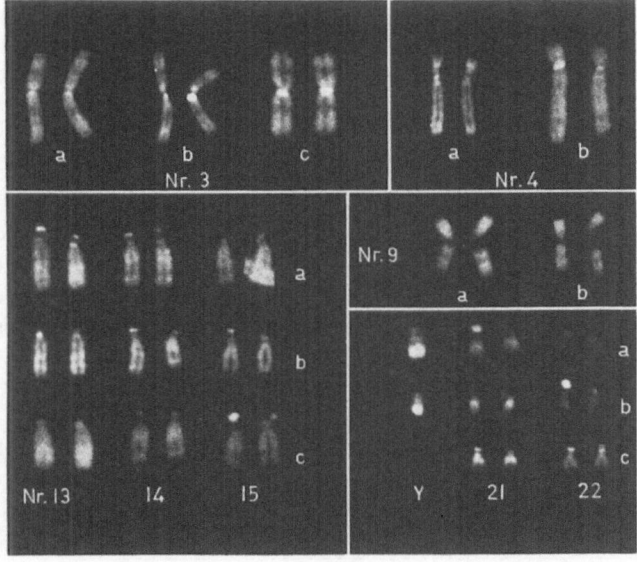

Fig. 29. Some of the variants of human chromosomes as seen after quinacrine staining. (From SCHNEDL 1974b)

Intense fluorescence is, however, quite easy to differentiate from pale fluorescence. The frequencies of the occurrence of such intense fluorescence on the variable areas of the acrocentric chromosomes are given in Table 1 (after SCHNEDL, 1974). It can be seen, that the secondary constrictions (the satellite stalks), although they may be of different length, never show any intense fluorescence. The centromeric region is quite variable in chromosome No. 13, in all other acrocentric chromosomes it shows intense fluorescence only seldom. Similar findings described in more detail have been published by MÜLLER and KLINGER (1974).

Table 1. Frequencies of Q-variabilities of the acrocentric chromosomes (59 cases from the Viennese population.) After SCHNEDL (1974)

Chromosome No.	Region	Frequency of strong fluorescence
13	Satellites	0.14
	Centromere	0.54
14	Satellites	0.22
	Short arm	0.01
15	Satellites	0.12
	Short arm	0.01
21	Satellites	0.28
	Short arm	0.01
22	Satellites	0.21
	Short arm	0.11

Prominent Q-Bands

The most striking variability (aside from the Y-chromosome and the acrocentric chromosomes) concerns *chromosome No. 3* (Fig. 29). The band in the long arm adjacent to the centromere (q11) is often intensely and sometimes brilliantly fluorescent (CASPERSSON et al., 1970). Sometimes somewhat weaker grades of fluorescence are observed, but it is always possible to decide whether or not if a chromosome No. 3 contains a strongly fluorescing spot in this region. On this basis it was found by several authors that the frequency of the occurrence is between 42% and 64% in normal white populations and 68% in a black population (SCHNEDL, 1971; PEARSON et al., 1973; MIKELSAAR et al., 1973, 1974; GERAEDTS and PEARSON, 1974; MÜLLER and KLINGER, 1974; see Table 2). MIKEL-

Table 2. Frequency of polymorphic chromosome No. 3 (brilliant Q-band near centromere) in different populations

Population	Estonia	Netherlands	Vienna	New York	
				white	black
Frequency	0.649	0.484	0.500	0.544	0.674
Authors	MIKELSAAR et al. (1974)	GERAEDTS and PEARSON (1974)	SCHNEDL (1971)	MÜLLER and KLINGER (1974)	

SAAR *et al.* (1974) and GERAEDTS and PEARSON (1974) reported also a significant sex difference.

A person can be positively or negatively homozygous, or heterozygous for this fluorescent spot. The distribution of heterozygous and homozygous conditions corresponds to the expected values. The condition is found to be the same in all cells of an individual and family studies show a clear Mendelian heredity (SCHNEDL, 1971).

Another variable Q-band is the centromeric region of *chromosome No. 4.* It is, however, of lesser intensity and much smaller in size than that on chromosome No. 3 (Fig. 29). It is therefore not possible to estimate its frequency in the population.

Y-Chromosome

The variation in length of the Y-chromosome has been thoroughly studied by several authors (e.g. COHEN *et al.*, 1966; UNNERUS *et al.*, 1967; COURT BROWN, 1967). With the aid of the fine banding techniques, particularly the quinacrine

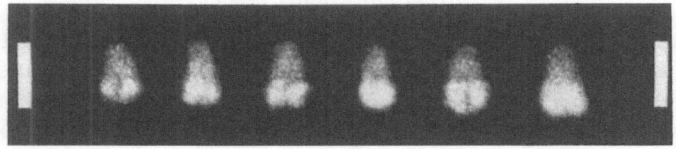

Fig. 30. Variants of the Y-chromosome as seen after quinacrine staining. (From SCHNEDL, 1971d)

fluorescence (Fig. 30), it can be shown that the distal part of the long arm, q12, which brilliantly fluoresces, is of variable length (BOBROW *et al.*, 1971; SCHNEDL, 1971). Some variations exist also in the non fluorescing proximal part of the long arm, q11 (SCHNEDL, 1971).

Measurements of the total length of the Y-chromosome revealed a Gaussian distribution within a given population (UNNERUS, 1967) but indications of racial differences have been reported (COHEN *et al.*, 1966).

Theoretical Implications

It has already been mentioned that the known chromosomal variations are not strikingly expressed in the phenotype and hence cannot have important direct genetic functions. At least they are not conspicuously expressed. Not even variations in the length of the Y-chromosome are linked with any sign of lesser fertility. It may, however, emerge with more and detailed studies on a larger scale, that certain chromosomal variations are connected with certain variants of the phenotype. The reported racial difference in Y-chromosome length (COHEN *et al.*, 1966) may point in this direction. Other important findings in this connection are the differences in the frequency of the fluorescing spot in the centromeric region of chromosome No. 3 between different populations. In Table 2 these frequencies in five population groups (four presumably predominantly caucasian, one black) are summarized. The evaluation of this chromosomal characteristic depends on a subjective judgement and may therefore be of only a limited value. The differences between white and black populations found by one group of authors (MÜLLER and KLINGER, 1974) are, however, remarkable. The sex difference, reported by MIKELSAAR et al. (1974) is less impressive, although the data are reported to be significant at the 5% level.

The chromosome polymorphism concerns mainly chromosomal regions which are heterochromatic and contain presumably higher amounts of repetitive DNA (see Chapter V). These regions are, at least partly, characterized by a deviating condensation mechanism. This may involve special proteins and protein bindings, eventually leading to a decreased genetic activity. On the other hand, a loss or gain of a certain number of DNA sequences, which are present in many copies from the beginning, may not influence the phenotype at all. This latter speculation may also explain how these variants came into existence. For instance, an unequal crossing over in meiosis could rather easily occur in such areas.

Practical Applications

With the quinacrine fluorescence technique different markers can be seen on autosomes and one marker on the Y-chromosome. They are, of course, all heritable features. Since the frequency of 2 markers is around 50% and that of some of the others over 20%, the probability for the differentiation between different karyotypes is quite high. The variations in karyotype may therefore be used for determining or excluding paternity in affiliation cases (LEISTI, 1971; SCHNEDL, 1974). SCHNEDL (1974) estimates that the possibility of excluding paternity exists in at least 70% of cases, provided that a good fluorescence technique is available.

It is of course also possible to decide whether a particular chromosome (which bears the marker region) is paternally or maternally derived. Such studies are important for the localization of genes on individual chromosomes. In this way it has already been shown by conventional staining methods that the locus for the Duffy blood group is on chromosome No. 1 (DONAHUE et al., 1968). Also the origin of the duplicated chromosome in trisomies could be shown in some cases (BREG et al., 1971; LICZNERSKI and LINDSTEN, 1972; ROBINSON, 1973).

If significant differences in the frequency distribution of the variabilities can be found between different populations or races (see Table 2), this could become quite important for ethnological studies.

V. Structural Differences Along the Chromosomes (Chromosome Banding)

1. Introduction

In the foregoing chapter different methods of bringing out structural differences along the chromosomes, the so called "bands", were described as well as their practical implications for identifying all the chromosomes. All these chromosome bands have been demonstrated in many different species. Practically every sort of mitotic chromosome, treated in an appropriate manner, displays some type of banding. Structural differences along the chromosomes can therefore be regarded as a general principle.

In this chapter mechanisms involved in banding of chromosomes and the significance of the bands will be discussed.

The cause of the appearance of chromosome bands may be manifold. During the last years, however, four factors have been found to be of particular importance:

The occurrence of repetitive DNA;

Differences in the base composition of DNA;

Differences in the protein components;

Differences in the degree of packing of the DNA or the DNP-complex.

These factors may of course be mutually related. It is for instance conceivable that differences in the protein component may lead to packing differences of the DNA and that a difference in base composition, or the occurrence of repetitive DNA, may be in turn connected with a difference in protein components. Moreover, all these factors may influence the disposition of the DNA to denature and reassociate under certain conditions, which may be an important factor in producing chromosome bands.

Some of the facts relevant for an understanding of the occurrence of a banding pattern in chromosomes are briefly summarized here. For detailed information and discussions on the underlying chemical mechanisms other reviews should be consulted (e.g. BRITTEN and KOHNE, 1968; WALKER, 1971; COMINGS, 1972; PARDUE and GALL, 1972).

2. Repetitive DNA

When DNA is centrifuged in cesium chloride gradients, the greatest part of it forms one main peak of buoyant density, but small peaks of varying density may also be observed (KIT, 1961; SZYBALSKI, 1961). These small peaks correspond to DNA-fractions which differ in their base composition from the main bulk of DNA. Fractions rich in guanine and cytosine (GC) have a higher buoyant density, whereas fractions rich in adenine and thymidine (AT) have a lower buoyant density. These deviating DNA-fractions are called "DNA-satellites" (Fig. 31). The expression "satellite" in this connection should not be mistaken for the small chromosome parts at the ends of the short arms of the acrocentric chromosomes.

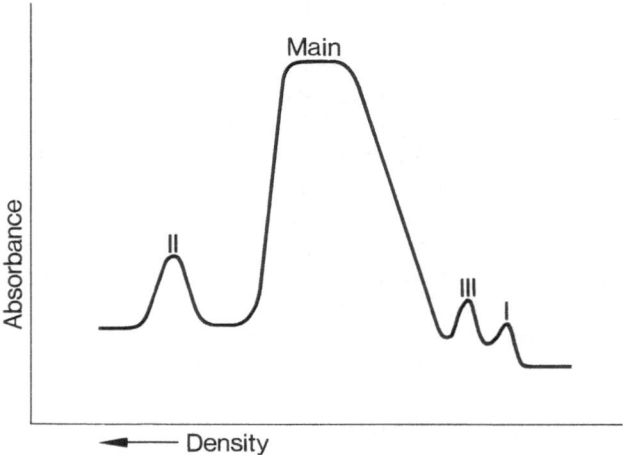

Fig. 31. Microdensitometric tracing of human native DNA centrifuged to equilibrium in Ag^+–Cs_2SO_4. Satellite DNA's are indicated by roman numbers. Sat. IV is very close to the main band on its light side. (After CORNEO et al., 1973)

In order to form a satellite, a DNA-fraction must consist of sequences which all deviate in the same direction. It is conceivable that in particular deviating sequences which are alike or similar in their base composition will form a satellite. It has been shown that satellite DNAs are indeed made up of identically repeated sequences (WARING and BRITTEN, 1966). When satellite DNA is dissociated into single strands (by heat or alkali treatment) it renatures very fast when compared to non repetitive DNA. WARING and BRITTEN (1966) found that mouse satellite DNA may contain as many as one million identical and repeated sequences. SOUTHERN (1970) reported the length of one of the repeated sequences in guinea pig satellite DNA to be as short as 8–13 bases. Different species have different and characteristic DNA satellites (FLAMM et al., 1969; HENNIG and WALKER, 1970; ARRIGHI et al., 1970; YUNIS and YASMINEH, 1970).

In *human DNA* at least four satellite fractions have been described so far (CORNEO et al., 1968; SAUNDERS et al., 1972 a and b; JONES et al., 1973a; CORNEO et al., 1973; ARRIGHI and SAUNDERS, 1973). Some of their characteristics are given in Table 3.

Table 3. CsCl and Ag^--Cs_2SO_4 densities of human satellite DNAs (G/ml) after CORNEO *et al.* (1973)

	Sat. DNA I	Sat. DNA II	Sat. DNA III	Sat. DNA IV
Buoyant density in neutral CsCl	1.687	1.693	1.696	1.700
Position in Ag^+-Cs_2SO_4	light	heavy	light	light
% of total DNA	0.5	2.0	1.5	2.0

DNA-satellite IV of CORNEO *et al.* (1972) is probably identical with the satellite described by SAUNDERS *et al.* (1972). It is to be expected that with refined methods even more satellites could well be isolated.

Of course, repetitive DNA which does not differ in buoyant density from the main band DNA cannot be detected by density gradient centrifugation. Other methods of isolating repetitive DNAs by means of their higher rate of reassociation have been developed. In general, 3 fractions can be isolated, highly repetitive, moderately repetitive and nonrepetitive DNA (BRITTEN and KOHNE, 1968). COR-NEO *et al.* (1973) found in human DNA two components within the CsCl-main band, called α and β components, which reassociate at a higher rate than the main bulk of DNA. These, in part moderately repetitious fractions, comprise about 17% of the total genome.

ARRIGHI and SAUNDERS (1973) concluded from reassociation studies, that as much as 30–35% of the human genome is composed of repetitive DNA. About half of this is highly repetitive (Cot* values from 0 to 0.05) the other half moderately repetitive (Cot values from 0.05 to 50). This means that the known satellite DNAs, which comprise only about 6% of the total DNA, correspond to less than half of the highly repetitive DNA present in the human genome. Thus highly repetitive DNA is included also within the main bulk DNA when investigated by density gradient centrifugation.

Clusters of repetitive sequences were also demonstrated in isolated eukaryotic DNA by their ability to form rings (THOMAS *et al.*, 1970). From such studies it was concluded, that in *Drosophila* about 50% of the genome consist of repetitive sequences which are concentrated in small regions, called g-regions (LEE and THOMAS, 1973). Such g-regions have an average length (of the DNA molecule) of 5 µm and their number equals roughly the number of cross bands seen micro-scopically in the giant chromosomes. Th g-regions in *Drosophila*, in *Necturus* and in the mouse are possibly composed of purely tandemly -repeating sequences or perhaps repeated sequences are spaced by non repetitive sequences (PYERITZ and THOMAS, 1973; BICK *et al.*, 1973; THOMAS *et al.*, 1973). Thus presumably also in man the amount of repetitive DNA is higher than has been estimated by the reassociation techniques. Also the possibility should be taken in considera-tion that the g-regions of THOMAS and coworkers (c.f.) correspond to the very fine bands in prophase- and premature condensed chromosomes which finally form the G-bands of metaphase chromosomes.

---

* Cot value: Co=initial concentration of DNA in moles/liter; t=reassociation time in seconds. The lower the reassociation time, the higher is the repetitiveness of the DNA.

3. Cytological Localization of Repetitive DNA

Localization of isolated DNA fractions in the chromosomes in cytological prepa-rations *in situ* has been achieved by the following procedure (PARDUE and GALL, 1970; JONES, 1970): The DNA fraction is used as a template for RNA which is synthetized *in vitro* from labelled precursors with the aid of RNA-polymerase. This labelled RNA is then hybridized with the single stranded DNA of the chromosomes *in situ*. To prepare the single stranded chromosomal DNA, chromo-some preparations are treated by heat and alkali to denature the DNA. The labelled RNA will only hybridize with those chromosomal regions which contain the corresponding DNA-fraction, and these regions can then be demonstrated by autoradiography. In this way it has been shown that in the mouse satellite DNA is localized at the centromeric regions (PARDUE and GALL, 1970; JONES, 1970).

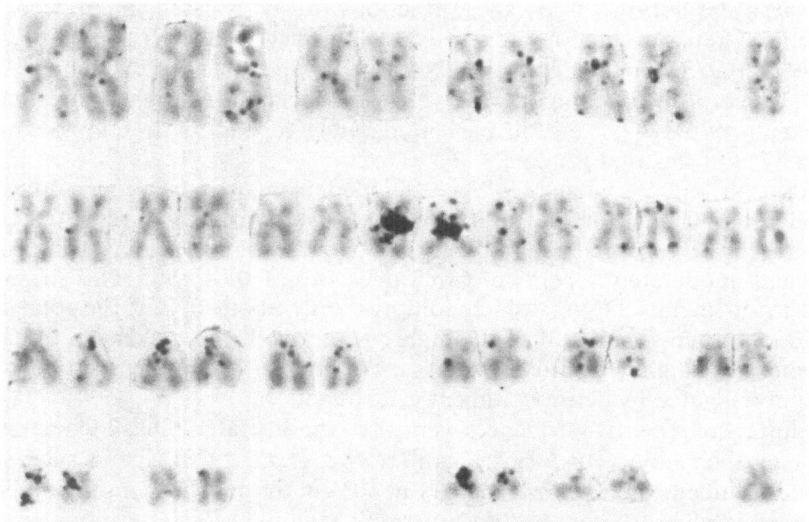

Fig. 32. Karyotype of a human male cell arranged from a metaphase showing the locations of hybridisation with labelled satellite III cRNA. (From JONES *et al.*, 1973)

In human chromosomes satellite DNAs have been also localized to some extent. Satellite II which is relatively A–T rich, is preferentially localized in the centromeric regions of chromosomes No. 1, 16, and to a slightly lesser degree 9 (CORNEO *et al.*, 1971; JONES and CORNEO, 1971; JONES *et al.*, 1973a). Satellite III is particularly enriched in the secondary constriction near the centromere of chromosome 9 (JONES *et al.*, 1973, see Fig. 32), the same region which stains preferentially with Giemsa at pH of above 11 (BOBROW *et al.*, 1972; GAGNÈ and LABERGE, 1972). Satellite I shows, in hybridisation experiments, a weak concentration in the centromeric regions of several chromosomes and a particular strong concentration in the distal part of the long arm of the Y-chromosome (JONES *et al.*, 1974). EVANS *et al.* (1974) reported that all four satellites are present in the Y chromosome in that part which shows brilliant fluorescence with quinacrine. In Fig. 33 a tentative map of the distribution of satellite DNAs in the human karyotype is given (JONES *et al.*, 1974). Satellite DNA is localized practically in all regions which are known to be constitutive heterochromatic.

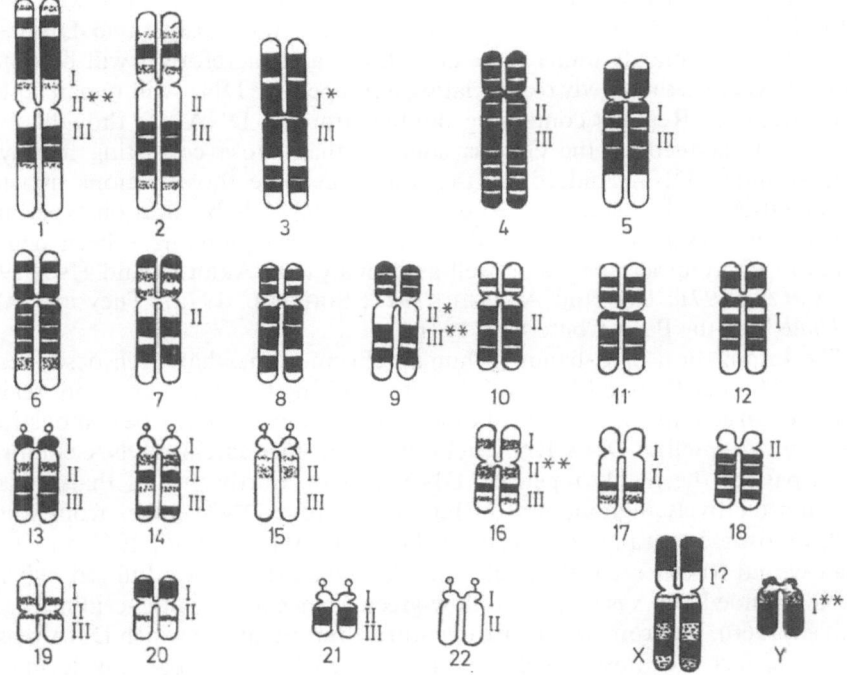

Fig. 33. Diagrammatic representation of the presently known locations of human satellite DNAs I, II and III. Locations with especially intense concentrations are noted with asteriks. (From JONES *et al.*, 1975)

Among the chromosomal regions containing repetitive DNA are also the nucleolar organizer regions coding for r-RNA (BIRNSTIEL *et al.*, 1968; LIMA DE FARIA *et al.*, 1969). Although nucleolar organizers are known to be situated in the secondary constrictions of the short arms of all 5 acrocentric chromosomes in man (OHNO *et al.*, 1961; HENDERSON *et al.*, 1972; BROSS and KRONE, 1973; EVANS *et al.*, 1974) no outstanding accumulation of satellite DNA has been found at these sites. This may be due to the low density of DNA in these regions. Another reason may be that presumably only about 60 tandemly repeated copies of the two genes for 18S and 28S r-RNA are present in each nucleolar organizer region in the human genome (OHNO, 1974). Furthermore it was demonstrated by MILLER and BEATTY (1969) that in *Triturus* the active ribosomal RNA coding genes are separated by non transcribing stretches of DNA. JONES *et al.* (1973), however, reported a certain tendency of satellite III to be localized in the nucleolar organizer regions. Satellite III is relatively G-C rich, as is known for ribosomal cistrons.

When chromosome preparations are treated in the same way as for an *in situ* DNA-RNA hybridization (by heat and alkali to denature the DNA, followed by a buffer solution to facilitate renaturation) a deep *staining with Giemsa* of chromosome regions containing repetitive DNA is observed (PARDUE and GALL, 1970). This has been interpreted in the following way: Repetitive DNA will reassociate very quickly under these conditions, and therefore it will be double stranded, whereas the slowly reassociating nonrepetitive DNA will remain chiefly single stranded. Regions containing double stranded DNA are thought to be more deeply stained by the Giemsa solution than those consisting mainly of single stranded DNA. Indeed, in the mouse genome those regions in which satellite DNA can be demonstrated by the DNA-RNA hybridization experiment (PARDUE and GALL, 1970) are deeply stained. Such regions have been demonstrated in many other species as well as in man (e.g. ARRIGHI and HSU, 1971; YUNIS *et al.*, 1971; HSU and ARRIGHI, 1971; SCHNEDL, 1972). They are called "*C-bands*" by the Paris Conference 1971.

The localization of C-bands in human chromosomes has been described in detail in Chapter IV (see Fig. 19). It is not surprising that there are more regions in the human genome stainable by the C-band procedure than can be demonstrated by the *in situ* satellite DNA-RNA hybridization, since satellite DNA comprises only a part of the highly repetitive DNA in man. Further proof that C-bands contain a relatively high amount of highly repetitive DNA comes from studies on chromosomes with acridine orange (DE LA CHAPELLE *et al.*, 1971 and 1973; BOBROW and MADAN, 1973). Single stranded DNA fluoresces dull red, whereas double stranded DNA is brightly yellow-green when stained with acridine orange (RIGLER, 1966). If chromosomes are denatured, the regions in which DNA reassociates very fast are those which are also stained by the C-banding technique. The distal half of the long arm of the Y-chromosome is the fastest and hence contains the highest amount of repetitive DNA, whereas the short arms of the acrocentric chromosomes are somewhat slower and contain probably less repetitive DNA.

There seems to be therefore no doubt that in general the C-band regions contain highly repetitive DNA. It must be remembered that some C-bands correspond to secondary constrictions (in the human karyotype the C-bands in chromosomes No. 1, 9, 16 and in the acrocentric chromosomes) and to the primary constrictions of the centromeres. That means that they concern regions in which the chromosomal material is *not* densely packed in metaphase. Another part of the C-bands corresponds to chromosomal regions which show a relatively high density (as the human Y chromosome in its distal part). Also, in other species, the C-bands may correspond to either particularly dense regions, e.g. the sex chromosomes of *Microtus agrestis* (PERA, 1970), or loose regions, e.g. the centromeric regions in chromosomes of bovidae (SCHNEDL and CZAKER, 1974) or the heterochromatin in the hedgehog (GROPP and NATARAJAN, 1972). The C-band regions are usually taken as equivalent with constitutive heterochromatin (see e.g. ARRIGHI and HSU, 1971). We should bear in mind, then, that constitutive heterochromatin is not necessarily strongly condensed in mitotic chromosomes (see also Chapter VIII).

It has been speculated that the *finer banding* of the chromosomes, the so called *"G-bands"* (see Fig. 21) are also due to the varying amounts of highly or moderately repetitive DNA, because some methods which produce them apply heat or alkali treatment as a first step and a neutral saline solution as a second step (SCHNEDL, 1971; DRETS and SHAW, 1971). G-bands are, however, also produced by saline solution treatment alone (e.g. SUMNER *et al.*, 1971) and also by proteolytic enzymes (DUTRILLAUX *et al.*, 1971; SEABRIGHT, 1972; WANG and FEDOROFF, 1972). They may even be visualized in standard preparations without any special pretreatment (McKAY, 1973; YUNIS and SANCHEZ, 1973). These latter findings do not directly contradict the hypothesis that the G-band regions contain higher amounts of repetitive DNA, but they show that, in addition, other components play an important role. SANCHEZ and YUNIS (1974) reported recently that repetitive DNA is indeed enriched in the same bands showing strong quinacrine fluorescence.

Since repetitive DNA is certainly enriched in constitutive heterochromatic regions, its *function* has often been only seen in connection with these.

It should be pointed out that several types of functioning genes are present in multiple copies. Besides the already mentioned genes for ribosomal RNA also genes for t-RNA must be present in multiple copies. OHNO (1974) points out that also promotor regions to which RNA polymerase is attached must be alike for each type of polymerase. He discusses also the possibility that identical operator regions exist in many copies and that cistrons coding for polypeptides showing internal homology such as tropocollagen may contain many tandemly repeated copies differing only slightly from each other.

The main amount of repetitive DNA can, however, be regarded as inactive "trivial" DNA. The constitutive heterochromatic regions containing highly repetitive DNA are doubtless genetically inactive or only a little active. But repetitive DNA (presumably the moderatly repetitive DNA) seems to be dispersed over all euchromatic regions, presumably enriched within the G-bands, as discussed above. OHNO (1974) regards trivial DNA as remnants from the process of evolution by gene duplication. He also provided strong arguments for the importance of the "doing nothing" of trivial DNA in order to decrease the mutational load.

4. Differences in Base Composition of DNA

In the foregoing paragraphs, it has been explained that similar base sequences are clustered at certain chromosomal regions, forming the so called repetitive DNA. If such sequences deviate from the main relation of GC/AT bases, they form satellites when density gradients are investigated. In man DNA satellites I and II are relatively AT-rich, whereas satellites III and IV are relatively rich in GC. Their chromosomal localization have already been discussed.

Several other findings indicate differences in base composition of DNA along the chromosomes. Important information is provided by studies on the quinacrine fluorescence dyes. WEISBLUM and DE HASETH (1972) as well as PACHMANN and RIGLER (1972) have shown that guanine reduces the fluorescence of both quinacrine mustard and quinacrine dihydrochloride bound to DNA. In regions relatively rich in GC, the quinacrine fluorescence is therefore quenched whereas AT rich regions will be brightly stained. This was confirmed by ELLISON and BARR (1972) who observed bright quinacrine fluorescence only in chromosomal regions (of *Samoaia leonensis*) which are AT-rich (as shown by differential ^{3}H-thymidine and ^{3}H-desoxycytidine labelling).

A difference of base composition was also very elegantly demonstrated by *in situ* localization of anti-A and anti-G immunoglobulins by immunofluorescence (DEV et al., 1972; SCHRECK et al., 1972). The bands which can be demonstrated by anti-A correspond to the bands seen by quinacrine fluorescence. Although the anti-G does not give such a clear picture, it is seen that this is localized at regions which are not stained or only weakly stained by quinacrine.

Further proof in the same direction comes from denaturation studies (DE LA CHAPELLE et al., 1973; BOBROW and MADAN, 1973). AT-rich DNA denatures at lower temperatures than GC-rich DNA (MANDEL and MARMUR, 1968). If chromosomes are gradually denatured by heat, those regions which are also brightly stained by quinacrine are the first which become single stranded as seen by the red acridine orange fluorescence staining (Fig. 34). The same mechanism seems to be responsible for the staining of the reverse banding (R-bands, DUTRILLAUX and LEJEUNE, 1971). In this method (incubation at 87° C in a relatively low concentrated ionic solution) the AT-rich regions will be selectively denatured, whereas GC-rich regions remain unchanged and stainable with Giemsa.

All this evidence leads toward the same conclusion, namely, that bands stained by quinacrine are relatively rich in AT, whereas the interband regions or R-bands are relatively rich in GC. There are, however, some exceptions: The secondary constrictions of chromosomes 1, 9, and 16 are known to be AT-rich, since they denature rapidly (DE LA CHAPELLE et al., 1973; BOBROW and MADAN, 1973) and since satellite II is localized there (CORNEO et al., 1973). However, these regions show no quinacrine fluorescence (see Fig. 26). WEISBLUM (1973) proposed that this may be explained by the spatial arrangement of guanine bases within the polynucleotide. His findings on quinacrine fluorescence of AT and GC polymers suggest that the degree of fluorescence enhancement is not only dependent on the AT/GC ratio, but also on the degree of interspersion of GC base pairs among AT base pairs. If the GC pairs are evenly distributed throughout the length of the polynucleotide, quenching should be particularly strong. Such a spatial arrangement can be expected in highly repetitive DNA. The proposal of WEISBLUM (1973) may also be valid for other cases in which a direct correspondence between AT-content and quinacrine fluorescence could not be established (e.g. BOSTOCK and CHRISTIE, 1974). Still further proof for the hypothesis that G-bands are rich in AT comes from studies with the dibenzimidazole derivate 33258-Hoechst. The degree of fluorescence which is produced by this compound depends on the AT-concentration in DNA. The banding pattern obtained with this compound corresponds to the Q-banding pattern (HILWIG and GROPP, 1973; WEISBLUM and HAENSSLER, 1974). LATT *et al.* (1974) were even able to show that the strongly fluorescing part of the human Y chromosome must be so rich in AT that labelling during S-phase with BrdU and successive fluorescence staining with dibenzimidazol results in a marked difference between the two sister chromatids.

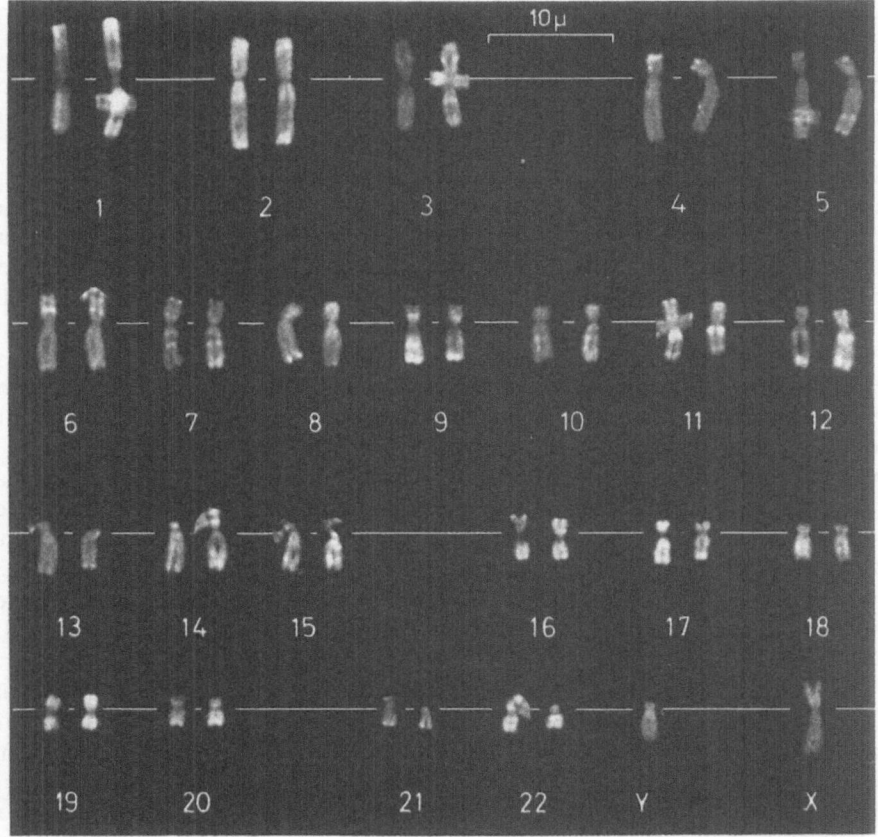

Fig. 34. Karyotype of a human male cell, stained with acridine orange after treatment with 95% formamide in SSC for 2 hours at 37° C, 0.25% formaldehyd being added during the last 3 min of incubation. Regions in which DNA is denatured appear red in the original preparation, and dull in the black and white photographe. (From CHAPELLE *et al.*, 1973)

Deoxy-5-methylcytidylic acid (5MeC). 5MeC has been found to be present as a minor component in chromosomal DNA (SHAPIRO and CHARGAFF, 1960). In some instances of mammalian DNA it has been found that 5MeC is concentrated in the satellite DNA fraction (e.g. in the mouse, SALOMON *et al.*, 1969). MILLER *et al.* (1974) were able to show by means of Anti-5MeC fluorescent antibodies that in mouse as well as in human chromosomes, 5MeC is concentrated at the constitutive heterochromatic regions of the chromosomes in both species.

5MeC is methylated *in situ* within the DNA molecule (SHEID *et al.*, 1968; LAWLEY *et al.*, 1972). Therefore in regions containing highly repetitive DNA, a high concentration of copies of recognition sequences for DNA methylase is present.

5. Differences in the Protein Components

The treatment of chromosomes with proteolytic enzymes (trypsin, pronase, pancreatin, chymotrypsin) results in a banding pattern which is generally similar to the Q- or G-banding pattern (e.g. DUTRILLAUX *et al.*, 1971; SEABRIGHT, 1972; WANG and FEDOROFF, 1972; FINAZ and DEGROUCHY, 1972; MÜLLER and ROSEN-KRANZ, 1972). This seems at first sight to be caused by a differential resistance of different chromosome regions to the action of proteolytic enzymes and hence may reflect differences in the protein components. It has been claimed, however, that inactivated trypsin solutions without any peptidase and esterase activity left are also able to produce chromosome bandings (SEHESTED, 1973). The action of trypsin seems therefore to be independent of its proteolytic property. McKAY (1973) suggested that proteolytic enzymes act by a removal of a cationic component which leads first to a swelling of the chromosome. When divalent cations are added, a reconstitution of chromosome morphology takes place which is more pronounced in the G-banding regions. The Giemsa solution itself may act in this way, because of its polyvalent cationic nature. It has indeed been shown that many other agents which act by removal of cations, such as detergents, potassium permanganate and urea, may produce chromosome banding (see e.g. UTAKOJI, 1972; KATO and YOSIDA, 1972). The chromosome banding effected by proteolytic enzymes may therefore reflect differences in the degree of chromatin packing.

LENG and FELSENFELD (1966) and CLARK and FELSENFELD (1972) found that AT-rich DNA is associated with lysine-rich histones and GC-rich DNA with arginine-rich histones. Since AT-rich and GC-rich regions do exist in alternating bands (demonstrable e.g. as Q-bands) lysine and arginine-rich histones should also be distributed in the same banding patterns. This is of particular interest in connection with the findings of LITTAU *et al.* (1965), that lysine-rich histones cause chromatin to condense, whereas arginine-rich histones prevent chromatin condensations. The Q-bands, which are with a few exceptions the same as the G-bands, would therefore represent regions which are AT-rich and which are at the same time more condensed. The importance of histone for the maintenance of chromosome structure after different treatments was also demonstrated by MEISNER *et al.* (1973).

RODMAN and TAHILIANI (1973) concluded from differential acid hydrolysis studies on mouse chromosomes, that dense bands may result from an association of a specific class of histones to DNA. On the other hand, GROSSEN (1973) found no banding patterns in chromosomes stained for histones by the fast green method. With this method, however, a differentiation between arginine- and lysine-rich histones is not possible.

YAMASAKI (1973) and GREILHUBER (1973) found that in plant chromosomes the constitutive heterochromatin behaves differential against HCl hydrolysis at different temperatures producing the so called Hy-banding pattern. Both authors favour the hypothesis that differences in non-histones cause the differential stability against HCl.

COMINGS et al. (1973) and VOGEL et al. (1974) showed that with the usual methods of producing a banding pattern, most of the histones are removed. They concluded therefore that histones cannot be responsible for the differential stainability. This conclusion may not be valid on all occasions and particularly not for chromosomes in vivo. As was pointed out above, a differential amount of histones, and of different types of histones (e.g. arginine-rich and lysine-rich) in different regions, may be responsible for differences in the degree of packing of chromatin or, in other words, in the density of DNA. As is shown in the next paragraph, packing differences exist without doubt along the chromosomes.

COMINGS et al. (1973, 1974), RODMAN and TAHILIANI (1973), MEISNER et al. (1974), VOGEL et al. (1974) and others, favour the hypothesis that the treatments necessary to produce bands themselves (above all the fixation in acetic acid-alcohol, the alkali, and the heat treatments) lead to a differential regional association of acid proteins to the DNA and in this way to differential increase in chromosomal condensation.

HSU et al. (1973) reported a banding pattern in chromosomes of cells which were treated with Actinomycin D (AMD) or with Ethidium bromide (ED) 2–6 hours before harvesting and suggested that these compounds bind to DNA at specific sites interfering with proteins necessary for condensing the chromosomes in mitosis. The observations of HSU et al. (1973) can presumably be better interpreted as a late labelling pattern, made visible by the interference of the Giemsa dye with the AMD or the ED which is incorporated into the late DNA synthethizing chromosome regions, similar to the effect of BrdU.

6. Packing Differences

In the foregoing paragraph it has been shown that the effect of proteolytic enzymes as well as the speculations on lysine-rich and arginine-rich regions, may reflect packing differences of the DNA or the DNA-protein complex. There are other and more direct observations indicating that most of the "bands" seen by many different methods are also connected with regions of a tighter packing of chromatin. The expression "chromatin" is used here meaning simply the material which builds up a chromosome. It has already been mentioned that the C-bands are heterochromatic regions. These regions are, however, not uniformly more tightly packed than euchromatic regions. Some (e.g. on chromosome 1, 9, and 16) are seen as secondary constrictions in metaphase.

McKAY (1973) reported that bands corresponding to the well known G-bands in acetic acid-alcohol fixed mouse chromosomes can be observed without any further treatment by phase contrast microscopy and ultraviolet microscopy (Fig. 35). YUNIS and SANCHEZ (1973) published similar findings, reporting even

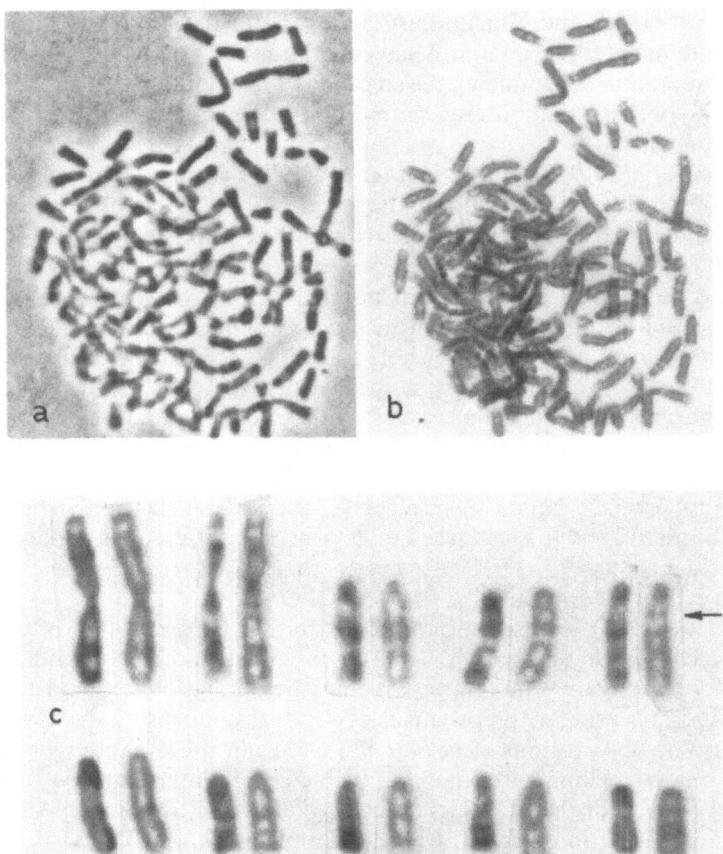

Fig. 35a–c. A metaphase from a mouse L-M cell line. In (a) before G-band staining in the phase contrast, in (b) after G-band staining in bright field. In (c) the same chromosomes before and after G-band staining. A very close similarity between the phase-contrast and the G-banding pattern exists. (From McKay, 1973)

a distinct banding pattern in human chromosomes after Feulgen staining, quite comparable to the Q- or G-banding pattern. The same was demonstrated in mouse chromosomes by Rodman and Tahiliani (1973). It seems to be clear that the uniform Feulgen staining of chromosomes, generally assumed up to the present, is not quite correct. Since the factor of an eventually differential denaturation by the hypotonic treatment and the acetic acid-alcohol fixation of standard chromosome preparations is, if any, quite small, the phase contrast and the Feulgen bands must reflect differences in chromatin density and in DNA content.

The most impressive proof for differences in the degree of chromatin packing comes from electron microscopy. Bahr and Golomb (1971) and Bahr et al. (1973) were able to demonstrate thickenings along the chromatin in preparations of chromosomes spread on a water surface and dried by the critical point method. Although this method leads presumably to a rather strong distortion of the original structure, these thickenings are very clear and their position corresponds to the Giemsa bands in many instances (Figs. 36 and 37). The same sort of strongly condensed bandlike regions can be seen in many earlier published electron microscopic photographs of whole mount chromosomes. As an example our own picture of a chromosome No. 13 is given in Fig. 38 (SCHWARZACHER and SCHNEDL, 1967).

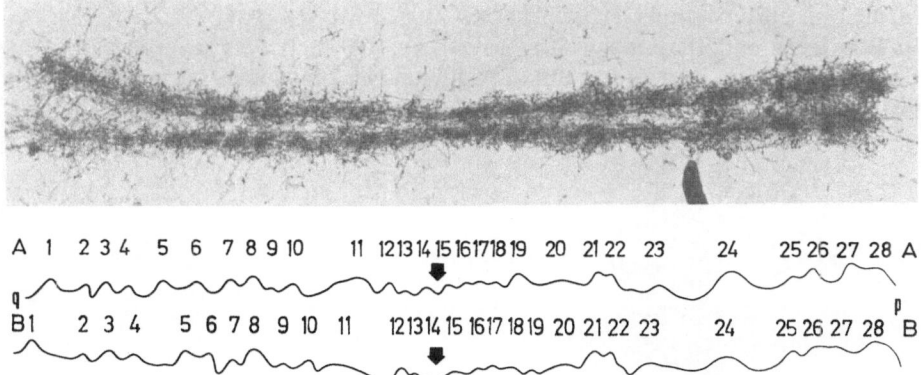

Fig. 36. Above an electron micrograph of a chromosome of group *A*, spread on water surface in the Langmuir trough, and dried at the critical temperature. Below densitometric tracing of both chromatids. The structural details in both chromatids are highly comparable. (From BAHR *et al.*, 1973)

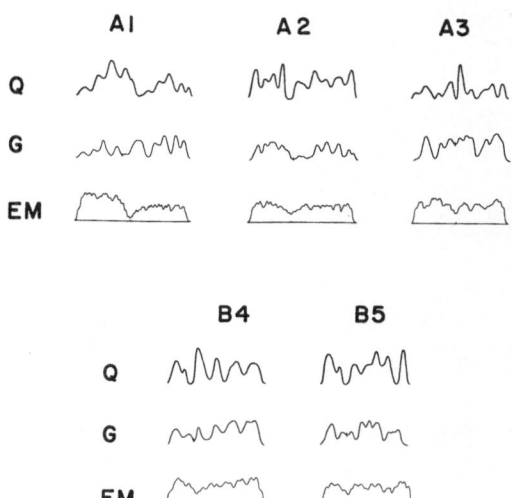

Fig. 37. Comparison of density tracings along the 5 largest human chromosomes after quinacrine staining (Q) Giemsa banding (G) and from electron micrographs after the method as described in Fig. 36 (EM). (From BAHR *et al.*, 1973)

The various treatments, leading to a pronounced banding pattern in the light microscope, enhance these density differences. Ross and GORMLEY (1973), McKAY (1973) and CERVENKA *et al.* (1973) showed by means of phase contrast and interference microscopy as well as microscopic studies after shadowing with heavy metals that the G-bands are much denser than the interband regions. The same can be observered in electron microscope pictures of whole mount chromosomes treated by banding techniques (CERVENKA *et al.*, 1973; Ross and GORMLEY, 1973; COMINGS *et al.*, 1973; RUZICKA and SCHWARZACHER, 1974, see Figs. 58 and 59).

Density differences along chromosomes can also be particularly well observed in meiotic chromosomes (HUNGERFORD, 1971; HUNGERFORD *et al.* 1971 a and b; BORDJAZE and PROKOFIEVA-BELGOVSKAYA, 1971). It was demonstrated that the bands of high density in human pachytene chromosomes correspond to the G-bands of mitotic chromosomes.

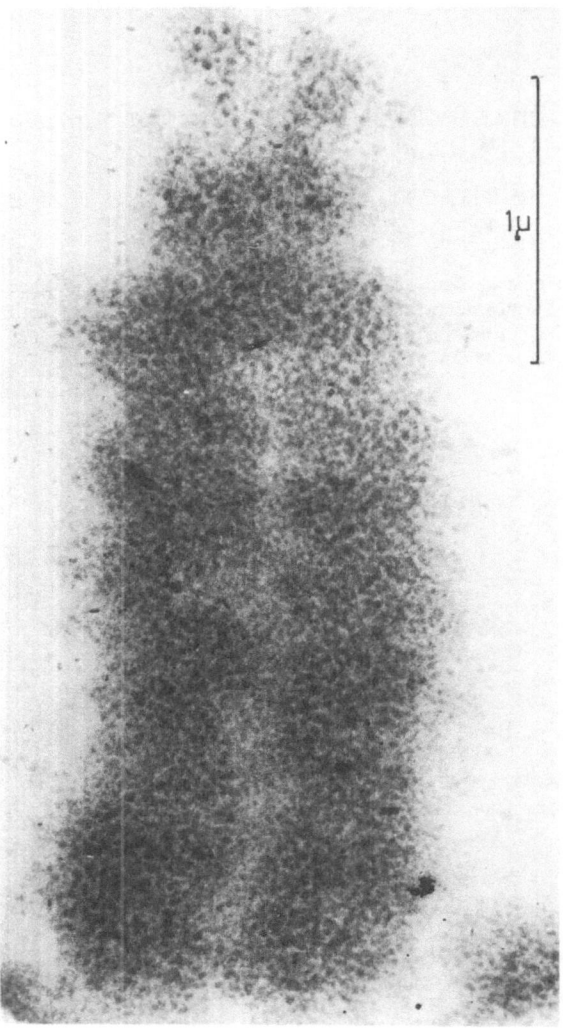

Fig. 38. A human chromosome No. 13, as seen in the electron microscope after standard chromosome preparation, HCl treatment and peeling off the slide with a collodium film. (After SCHWARZACHER and SCHNEDL, 1967)

7. The Binding of Giemsa to Chromosomal Components

The differential staining of chromosomes by Giemsa-solution after various pre-treatments may be better understood if the mechanism of the binding of the dye to chromatin is known. SUMNER and EVANS (1973) found that Giemsa is largely bound to DNA and not to proteins. MCKAY (1973) and ROSS and GORMLEY (1973) reported that the special treatment necessary to produce or enhance banding act first by an alteration of chromatin structure and then by a differential reconstitution. As McKAY (1973) suggests, the alteration is brought about by removal of cations and the reconstitution takes place on the addition of divalent cations. Giemsa itself may act reconstitutively because of its cationic nature. COMINGS et al. (1973), MEISNER et al. (1973) and VOGEL et al. (1974) have pointed out that due to the fixation and other treatments leading to banding, the protein-DNA complexes may be differentially altered and thus produce the staining differences.

SUMNER and EVANS (1973) believe that with Giemsa, a magenta compound is formed in situ which is attached to DNA molecules by hydrogen bonds. They suggest that the amount of Giemsa bound depends on the spatial arrangement of the binding sites of the chromatin. They have stated in addition that the binding of Giemsa is not affected by whether the DNA is double stranded or denatured.

In any case, these investigations indicate that the Giemsa stain is affected by the kind of packing or spatial arrangement of the DNA-protein complex.

8. Conclusions on the Mechanisms Involved in Banding

C-bands: There is little doubt that repetitive DNA is enriched in C-band regions. Also, packing differences seem to be important: the C-bands are believed to correspond to constitutive heterochromatin which was originally defined by a high density during interphase (HEITZ, 1928, 1929). As has been pointed out above, however, only the distal part of the Y-chromosome shows particularly dense chromatin packing, whereas most other C-regions correspond to primary or secondary constrictions. C-band regions also contain a particularly high concentration of methylated cytosine.

The increased readiness of repetitive DNA to reassociate may very well influence its stainability. C-band staining gives therefore no direct indication of density differences. The connection between special types of histone or non-histone proteins and C-band regions seems to be not quite clear.

Q-bands: Quinacrine dyes have a definite tendency to stain AT-rich regions. In addition, density differences along the chromosomes may enhance Q-banding. Moreover the differential rate of denaturation of AT- and GC-rich regions may underline the banding. Finally the possibility that Q-bands contain higher amounts of repetitive DNA than the interband regions may also play a role.

G-bands: The differential staining with Giemsa is probably caused either by a differential reconstitution of the chromosome after it has been distorted by the removal of cations, or by a differential alteration of the DNA-protein complex, or by both. Several factors, such as the relatively high density of the chromatin packing, the increased A-T content, the relatively high amount of repetitive DNA may lead to this behaviour as a result of treatments necessary to produce the G-bands.

R-bands: The lower degree of denaturation of the GC-rich R-bands due to the treatment by this method seems to be a major factor for the differential staining in this case.

To *summarize*—we find regions in the chromosomes containing high amounts of highly repetitive DNA. These are the C-bands, or the constitutive heterochromatic regions. They are in some instances connected with a very dense packing of chromatin, in others they are particularly loosely packed. Besides these, there are the Q- and G-bands which are relatively densely packed and contain moderate amounts of repetitive DNA. Most of these bands are rich in AT. Finally, the interband regions (corresponding also to the R-bands) are not densely packed and represent relatively GC-rich regions with the least amount of repetitive DNA. So far, not much about different protein components is known. The finding that the AT-rich bands are associated with lysine-rich histones, as well as the fact that histones are presumably removed to a high degree during preparation, and that differential binding of acid proteins to DNA may play a role, have to be borne in mind.

9. DNA-Replication Pattern

In mitotically dividing cells, the chromosomal DNA is doubled during a defined period in interphase, the so called synthesis or S-period. Before the S-period, the DNA content of the interphase nucleus is constant; this period is called the first gap period, or G_1-period. In the same way the DNA content remains constant after the end of the S-period until the chromosomes divide in mitosis; this period is called the second gap-period, or G_2-period (HOWARD and PELC, 1953). The S- and G_2-periods are of relatively similar and constant duration in many different tissues and species. The G_1-period is the one which differs most, and which is mainly responsible for the great differences in the duration of the cell cycle (i.e. the whole period between two mitoses) of different tissues.

In logarithmically dividing human fibroblast cultures, the duration of the S-period is 7–9 hours, in human lymphocytes of short term cultures from peripheral blood 12–15 hours. The duration of the G_2-period varies between 2–5 hours in both types of cells (e.g. BENDER and PRESCOTT, 1962; GRUMBACH et al., 1963; MOORHEAD and DEFENDI, 1963; SCHWARZACHER and SCHNEDL, 1965; BIANCHI and BIANCHI, 1965).

During the S-period not all chromosomes and chromosome regions replicate DNA synchronously. Notably, the heterochromatic regions begin and finish DNA synthesis later than the euchromatic regions (LIMA DE FARIA, 1959; TAYLOR, 1960; GRUMBACH et al., 1963). There are also finer differences in the timing of DNA replication in practically all chromosomes. Particularly towards the end of the S-period, a consistent pattern exists of chromosomal regions which are still synthesizing or which have already finished synthesis. This is called the "late DNA-replication pattern". SCHMID (1963) and GERMAN (1964b) provided the first descriptions of this late replication pattern in human chromosomes. The technique to make these patterns visible used until recently was autoradiography after incorporation of ^{3}H-thymidine into the synthetizing cells. ^{3}H-thymidine is added to a cell culture, and 5–6 hours later chromosome preparations are made. In this way, those chromosomal regions which incorporated the ^{3}H-thymidine during the last 1–3 hours of the S-period will be labelled (Fig. 39). It must be kept in mind, however, that the labelling pattern after ^{3}H-thymidine may also indicate chromosome regions particularly rich in AT.

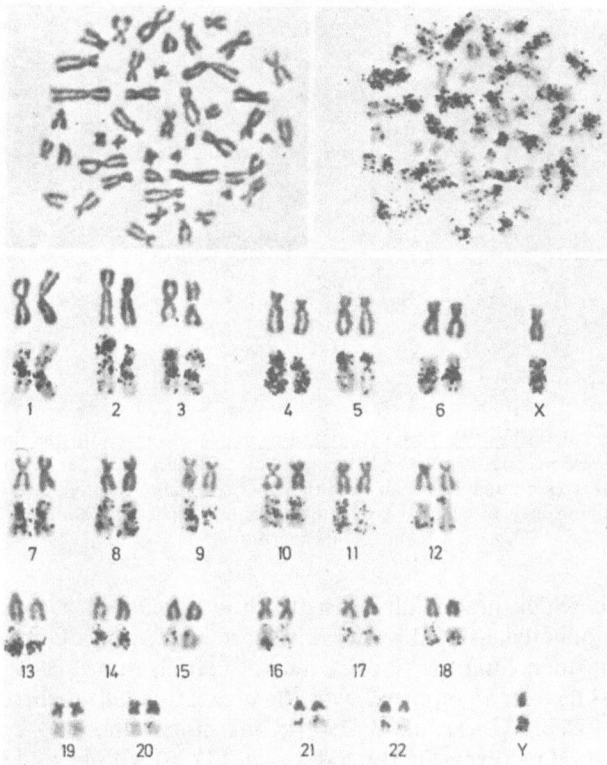

Fig. 39. Metaphase figure and karyotype of a cell labelled with ^{3}H-thymidine for the final 6 hours before preparation. Standard orcein stain and autoradiography for comparison, typical late DNA replicating pattern ("intermediate" stage). One chromosome No. 4 shows a deletion on the short arm. Lymphocyte culture, ^{3}H-TdR 0.3 µCi/ml, 1.9 Ci/mMol spec. act., AR stripping film for 10 days. ca. × 1000. (From PASSARGE, 1974)

ZAKHAROV *et al.* (1971, 1972, 1973, 1974) have developed a new methodical approach to make DNA replicating chromosome regions visible: If BrdU (Bromuridinedeoxyribose) is given in the proper concentration during the S-period it will be built in into those regions which are in the process of replicating.

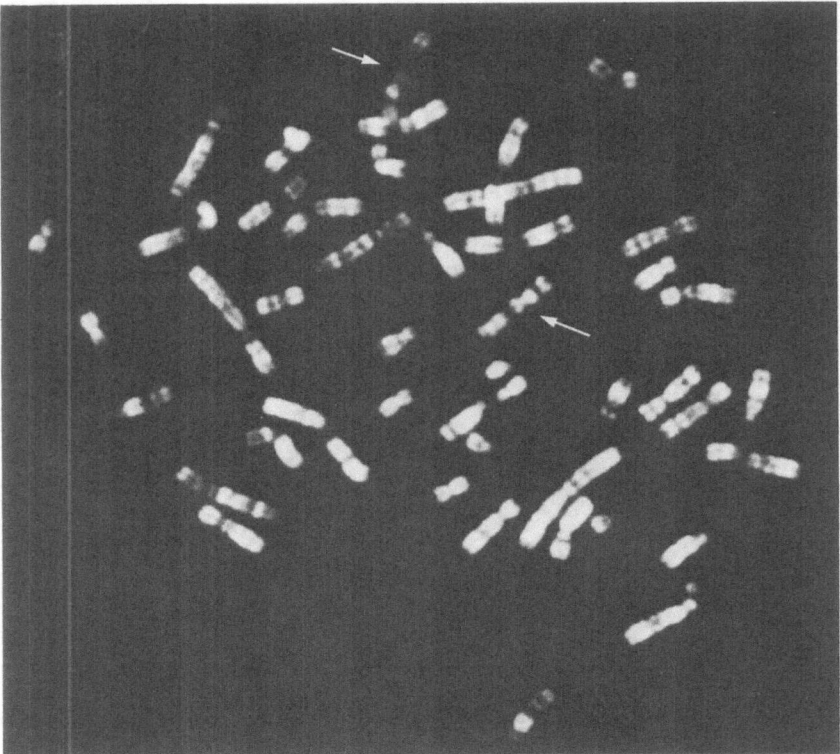

Fig. 40. Human female metaphase. BrdU (10 μg/ml medium) was given for the final 6 hours before preparation. Fluorescence staining with bis-benzimidazole (Hoechst 33258). The regions in which BrdU is built in fluoresce much less than the brightly fluorescing other regions. The arrows point to the two X-chromosomes, of which one (the upper) is late replicating and hence weaker stained.
Courtesy of M. Mikkelsen

DNA. Regions which have built in BrdU show a marked decondensation and thus appear elongated as well as weakly stained. The BrdU has furthermore the property of quenching the fluorescence of certain fluorescent dyes (Fig. 40). The best results have been reported with the benzimidazolfluochrome compound "33258" of Hoechst (LATT, 1973, 1974), but quinacrine and acridine orange may also give good pictures (DUTRILLAUX et al., 1973). GROPP and HILWIG (1973) found that the compound 33258-Hoechst leads to a decondensation of the hetero-chromatic regions when built in into chromosomes of living cells. It causes also a decrease of the fluorescence of these regions when they are again stained with 33258-Hoechst after fixation. Thus, this compound may perhaps directly be used to study late replication. A further technical improvement was achieved by PERRY and WOLFF (1974), WOLFF and PERRY (1974), and KORENBERG and FREEDLENDER (1974) who stained BrdU-treated cells first with the Hoechst-compound and afterwards by a Giemsa-banding method (Fig. 41). These findings suggest that differences in the binding of protein to BrdU-containing chromosome segments cause the differential staining with Giemsa (IKUSHIMA and WOLFF, 1974).

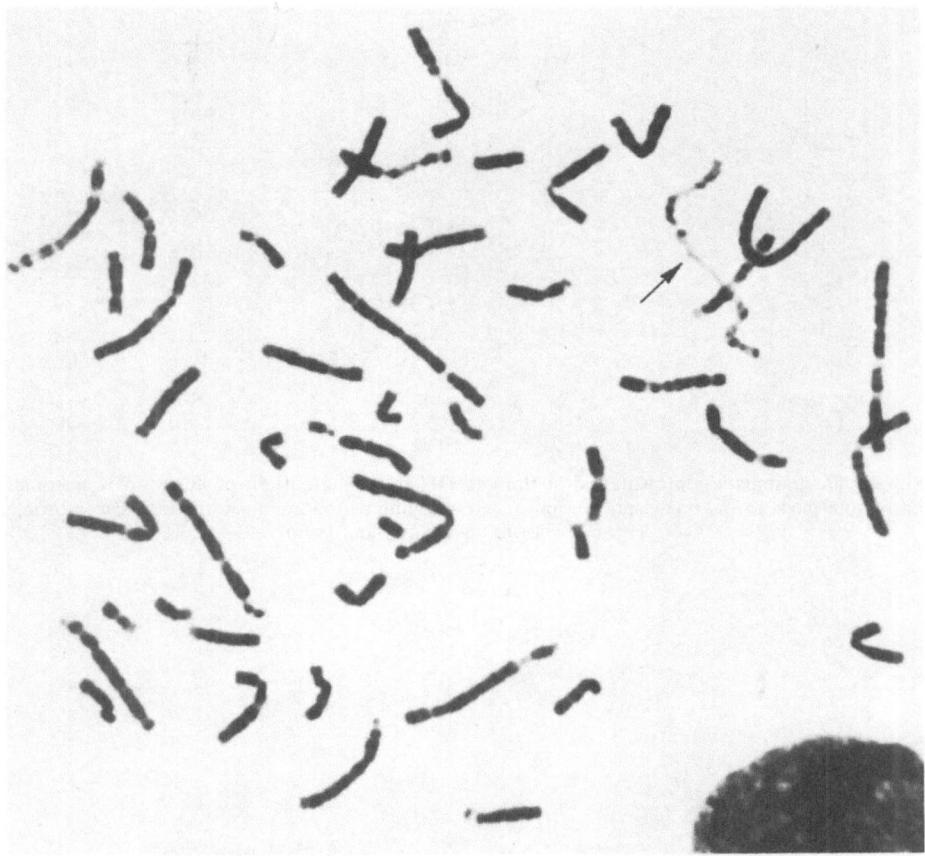

Fig. 41. Human female metaphase. Similar treatment with BrdU as in Fig. 40. Staining with Giemsa similar to the method of PERRY and WOLFF (1974). The thin weakly stained chromosome (arrow) is the late replicating X

The autoradiography of late labelling chromosomes was the only method to identify individual human chromosomes beyond pure morphological methods before the new banding techniques were introduced. Today this method and its new BrdU-variant still have their value in certain cases. For instance, one of the two X-chromosomes of a female cell replicates particularly late (Fig. 40). This is the genetically inactive X-chromosome which is facultatively heterochromatic (see chapter VII). Structural abnormalities or translocations involving this chromosome can be best investigated by the late labelling method (e.g. PASSARGE, 1974; MICKKELSEN, 1974).

Extensive descriptions of replication patterns in normal and abnormal human mitotic chromosomes exist (e.g. SCHMID, 1963; GERMAN, 1964; GIANELLI, 1970; MILLER, 1970; PASSARGE, 1974). Most investigators divide the end of the S-period into early, intermediate, and late stages following a recommendation by SCHMID (1963). In Fig. 42 a schematic representation of the late labelling pattern is given.

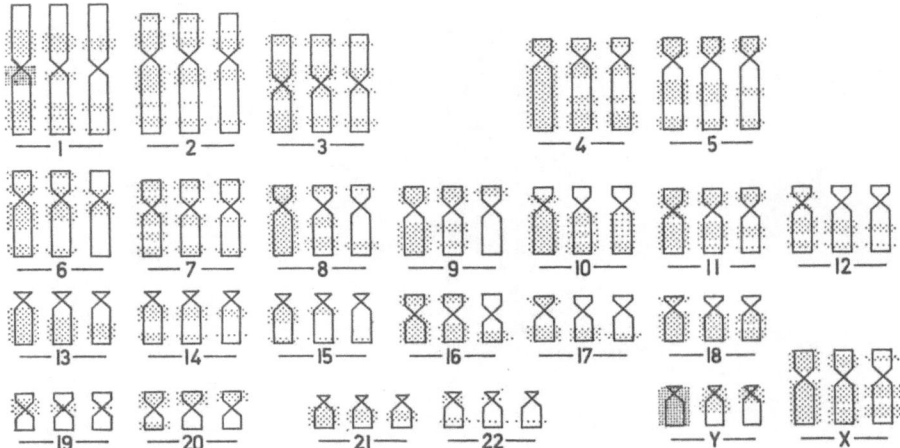

Fig. 42. Diagrammatic representation of the late DNA labeling pattern of human chromosomes. Early, intermediate and late stages within the late labelling period are given for each chromosome. (From CALDERON and SCHNEDL, 1973)

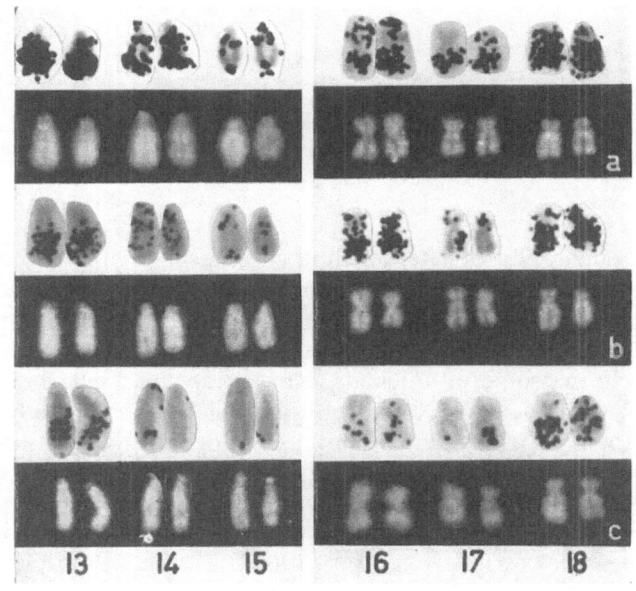

Fig. 43a–c. Comparison of late labelling pattern and quinacrine fluorescence for chromosomes 13–18. (From CALDERON and SCHNEDL, 1973)

Of particular interest are studies comparing the late DNA replication patterns with the C-, Q- and G-banding on the same cells (GANNER and EVANS, 1971; BREG et al., 1972; CALDERON and SCHNEDL, 1973). These studies, all done with the autoradiographic method, have revealed that late replicating regions and bands coincide in many cases. The resolution of the autoradiographic method is not fine enough to compare all the fine bands revealed by the quinacrine or Giemsa techniques. Detailed studies using the new BrdU-method have not been done so far. However, it has been shown that the C-bands are in general among the latest replicating zones and that also the G-bands and Q-bands correspond to relatively late replicating regions (Fig. 43). This adds an important information to our knowledge of the nature of the bands. The C-bands correspond to heterochromatic regions which have long been known to synthetize DNA relatively late during the S-period (LIMA DE FARIA, 1959). So this is what one would expect. There is no doubt about the late replication of constitutive and facultative heterochromatin. That these regions start and finish DNA replication definitely later than euchromatin has been shown not only by autoradiographic studies using ^{3}H-thymidine (e.g. PERA, 1968) but also by studying premature condensed chromosomes (SPERLING and RAO, 1974).

The relatively late replication of the Q- and G-bands again indicate that they are in some way similar to the C-bands although not so pronouncedly different from the rest of the chromosomal regions. However, no definite correlation between the amount of repetitive DNA, packing density, special protein bindings, and late replication can be made so far. It has already been mentioned that ^{3}H-thymidine incorporation may also indicate a relative richness of AT in the chromosomes. The same is of course true for BrdU, since uridine instead of thymidine is incorporated into DNA. Thus, the socalled "late"-labelling pattern may also reflect an AT-pattern. This may be one of the reasons that Q-bands and thymidine replicating patterns coincide. SCHNEDL (1973) has shown that ^{3}H-deoxycytidine can produce a somewhat similar late labelling pattern as ^{3}H-thymidine. The pattern obtained with ^{3}H-deoxycytidine is, however, in many cells less distinct than that with ^{3}H-thymidine, and shows differences in the secondary constrictions of chromosomes 1 and 16, which are known to be GC-rich. Therefore the thymidine labelling pattern does not necessarily reflect a late DNA synthetizing in all chromosome regions but may be modified by the relative AT-richness of certain bands.

10. Chromomeres and G-Bands

Special attention should be paid to a comparison of the bands in mitotic chromosomes with the well known bands or chromomeres of polytene interphase chromosomes (e.g. salivary gland giant chromosomes of insect larvae). BAHR et al. (1973) reported that in electron microscopic preparations of human *metaphase* chromosomes as many as 225 major and 450 minor dense bands can be seen per haploid set. This figure approaches the number of bands in the giant chromosomes of various Dipterans, which is between 2000 and 5000 (BEERMANN, 1972). In *inter-*

phase, the G-bands are conceivably much finer and much more numerous than in metaphase. G-banding methods applied to premature condensed chromosomes do indeed show a very fine banding pattern along the thin threadlike chromosomes (UNAKUL *et al.,* 1973; SCHWARZACHER *et al.,* 1974; RÖHME, 1974). The fine bands are of different sizes and are separated from each other by interband zones (Figs. 44 and 45). From electron micrographs (Figs. 63 and 64) it can be estimated that the number of bands in the human genome lays in the order of 10^4 to 10^5. Prophase chromosomes also show a much finer banding than metaphase chromosomes (Fig. 24). The finer bands obviously fuse during chromosome condensation to form the coarse bands of metaphase.

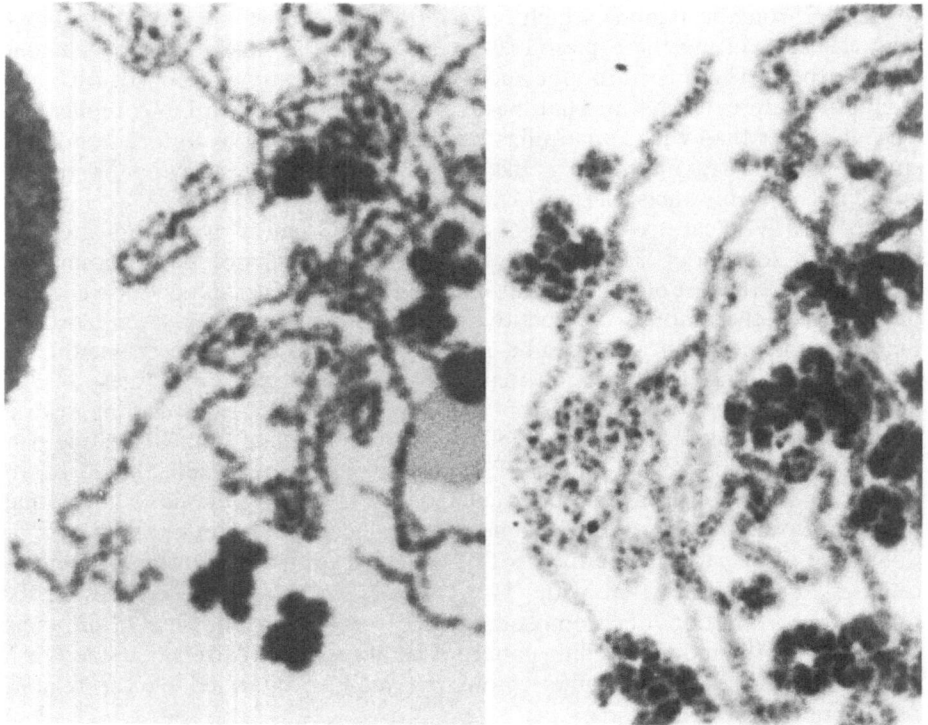

Fig. 44   Fig. 45

Figs. 44 and 45. Fused cells showing premature condensed G1-chromosomes. G-band staining. The still thin and elongated premature condensed chromosomes show a very fine banding pattern.
$\times 2500$

The G-bands may well correspond to chromomeres of giant chromosomes on the basis of the similar number and similar structural feactures. In this connection it is of high interest, that LEE and THOMAS (1973) found by studying ring formations of isolated DNA molecules of *Drosophila,* that the number of regions containing repetitious DNA sequences (called g-regions) is approximately equal to the number of microscopically visible bands in the giant chromosomes. This indicates that the so called g-regions (containing the repeated sequences, see

PYERITZ and THOMAS, 1973, and THOMAS *et al.*, 1973) correspond to the G-bands discussed in this chapter.

No direct comparison of the fine G-banding pattern of mitotic chromosomes with the chromomeres of giant chromosomes has been done so far in the same species. HSU (1972) has shown that in metaphase chromosomes of *Drosophila melanogaster* the C-banding method stains exactly the same regions which appear in giant chromosomes as heterochromatic in the classical sense.

11. Genetic Mapping in Human Chromosomes

The possibility to identify each chromosome and certain regions of chromosomes was the premise to assign gene loci to specific chromosome regions. A further great help was the possibility to use somatic cells in vitro for genetic studies. Besides the use of naturally occurring chromosomal variants and rearrangements the introduction of a system in which somatic segregation of chromosomes occurs regularly was of great importance. This was done by using somatic cell fusion *in vitro*.

Somatic cell fusion was carefully studied by BARSKI *et al.* (1960) and EPHRUSSI and SORIEUL (1961). A method to induce it by the help of inactivated Sendai virus was developed by HARRIS and WATKINS (1965). It is possible to produce intraspecific as well as interspecific cell hybrids by this method (see HARRIS, 1970). If two cells fuse the two cell nuclei will combine to form a heterokaryon which then contains both parental genomes. The important fact is that heterokaryons have a tendency to lose chromosomes. When biochemical markers, which are expressed in somatic cells *in vitro,* are used, such markers can be assigned to specific chromosomes if it is possible to correlate e.g. their disappearance in a cell clone with the loss of the chromosome in question. A particularly well suited system are man/mouse hybrid cells, because the human chromosomes segregate quite constantly during the first generations after the fusion, and relatively constant clones with a varying number of human chromosomes develop (see RUDDLE, 1970). A description of the methods applied to correlate biochemical markers to specific chromosome segments are beyond the scope of this article. The reader is referred to the reviews in this field (e.g. RUDDLE, 1972; KUCHERLA-PATI *et al.*, 1974). A comprehensive up to date summary on the localisation of human genes on the individual chromosomes is given in the New Haven Conference (1973).

VI. Fine Structure of Chromosomes

1. Introduction

Light microscopic studies of chromosomes since the time of the discovery of mitosis in 1878 have shown that the chromosomes of practically all eukaryotic species look very similar. It is therefore widely assumed, and in fact it has been shown for a great number of different and quite unrelated species, that the finer structural elements of chromosomes as well as their morphological organisation are similar. Despite the fact that this uniformity allows the study of the most favorable species our knowledge of the fine structure of chromosomes is still incomplete. One major reason for this is certainly the problem of artefacts, which arise with every sort of electron microscopic investigation. From ultrathin sections, the method which creates probably the least amount of artefacts, it is very hard to construct three dimensional pictures because of the very small size of the structural elements found in chromosomes. On the other hand, the methods of preparing chromosomes *in toto* are all connected with rather crude procedures causing many artefacts.

Nevertheless it became clear from the first electron microscopic studies that fibrils are the main constituents of chromosomes (NEBEL, 1958; KAUFMANN *et al.*, 1960; RIS, 1961). Earlier light microscopic studies had already shown that the metaphase chromosome results from a coiling of the thin threadlike prophase chromosome (e.g. DARLINGTON, 1955). It was often postulated, therefore, that a thin prophase chromatid is also the result of the coiling of a still finer fibril, and that in general a chromatid is built up of one very fine fibril representing the deoxyribonucleoprotein molecule coiled and supercoiled several times. On the other hand several publications including very recent ones, seemed to suggest that two or more fibrils run parallel to form a chromatid (e.g. RIS, 1961, 1966; OSGOOD *et al.*, 1964; WOLFE, 1965; GALL, 1966; SORSA, 1973a; STUBBLE-FIELD, 1973). Thus there are two main problems to discuss:

1. The fine structure of the fibrils and their possible relation to the DNA molecule.

2. The arrangement of the fibrils within the chromosome, particularly the question as to whether one or more fibrils build up a chromatid.

2. Structure of the Fibrils

It has been shown by a variety of methods of preparations that chromosomes are composed of fibrils. Depending on the method however, the dimensions and the fine structure of the fibrils have been differently reported.

Ultrathin Sections

Section preparations have the great advantage that because of the possibility of an optimal fixation *in situ* without any deforming pretreatment, the fine structure of the fibrils is preserved with a minimum of artefacts. Two classes of fibrils have been consistently reported by many investigators using material from quite different species: ca. 20–30 Å thick fibrils, and ca. 100 Å thick ones.

The elementary fibril of 100 Å diameter. Fibrils of an average thickness of 100 Å (ranging from about 80 to 150 Å) have been found by practically everybody who investigated thin sections of *in situ* embedded chromosomes or cell nuclei with the electron microscope. This fibril may be regarded as the main structural element of chromosomes and can be called the elementary fibril. It has been reported, for instance, in *Gryllus argentinus* (WETTSTEIN and SOTELO, 1964), *Triturus viridescens, Rana pipiens* (WOLFE and GRIM, 1967), cultured HeLa cells (ROBBINS and GONATAS, 1964) as well as in cultured euploid human cells (BARNICOT and HUXLEY, 1965; SCHWARZACHER and SCHNEDL, 1967; LAMPERT, 1969; LAMPERT and LAMPERT, 1970). Figs. 46 and 47 show sections through glutaraldehyde fixed and epon embedded human chromosomes. Fibrils of 100–120 Å thickness can readily be seen.

Fine fibrils as components of the elementary 100 Å fibril. Fibrils of a diameter of 20–30 Å have also been reported by many investigators in thin electron microscopic section preparations (e.g. RIS, 1961; COLEMAN and MOSES, 1964; BRINKLEY and BRYAN, 1964; WETTSTEIN and SOTELO, 1965; SCHWARZACHER and SCHNEDL, 1969; LAMPERT and LAMPERT, 1970). Very clear pictures showing that in chromosomes of *Gryllus argentinus* an approximately 20 Å thick fibril is folded into an elementary fibril measuring 100–120 Å in diameter were published by WETTSTEIN and SOTELO (1965). In human metaphase chromosomes similar findings were made by SCHWARZACHER and SCHNEDL (1967). In Fig. 48a high magnification of a glutaraldehyde fixed, epon embedded and ultrathin sectioned chromosome demonstrates that the 100–120 Å fibrils are composed of a 20–30 Å fibril laid in rather irregular foldings. No indication is found that more than one such fibril forms one elementary fibril.

The 20–30 Å fibril is within the dimension of the DNA double helix molecule, which has a diameter of 22 Å. The measurements in electron microscopic photographs are in most cases not sufficiently exact to determine with certainty the diameter of the fine fibrils within a range of about ± 10 Å. It seems difficult therefore to decide, whether the fine fibrils represent the bare DNA molecule or a DNA-protein fibril. LAMPERT and LAMPERT (1970) suggested that the 30 Å fibre is the basic DNA-protein fibre which is coiled to form a 70–100 Å thick fibril (the elementary fibril). They base this suggestion on the structure of the thicker fibril which has sometimes the appearance of a hollow tube. The pictures published by WETTSTEIN and SOTELO (1965) as well as our own (Fig. 48) are more suggestive of a somewhat irregular folding of the bare DNA molecule. They may even be interpreted in terms of the model proposed by KORNBERG (1974) and by BALDWIN *et al.* (1975), consisting of the DNA molecule winding around centrally situated globules of histones (see Fig. 61a).

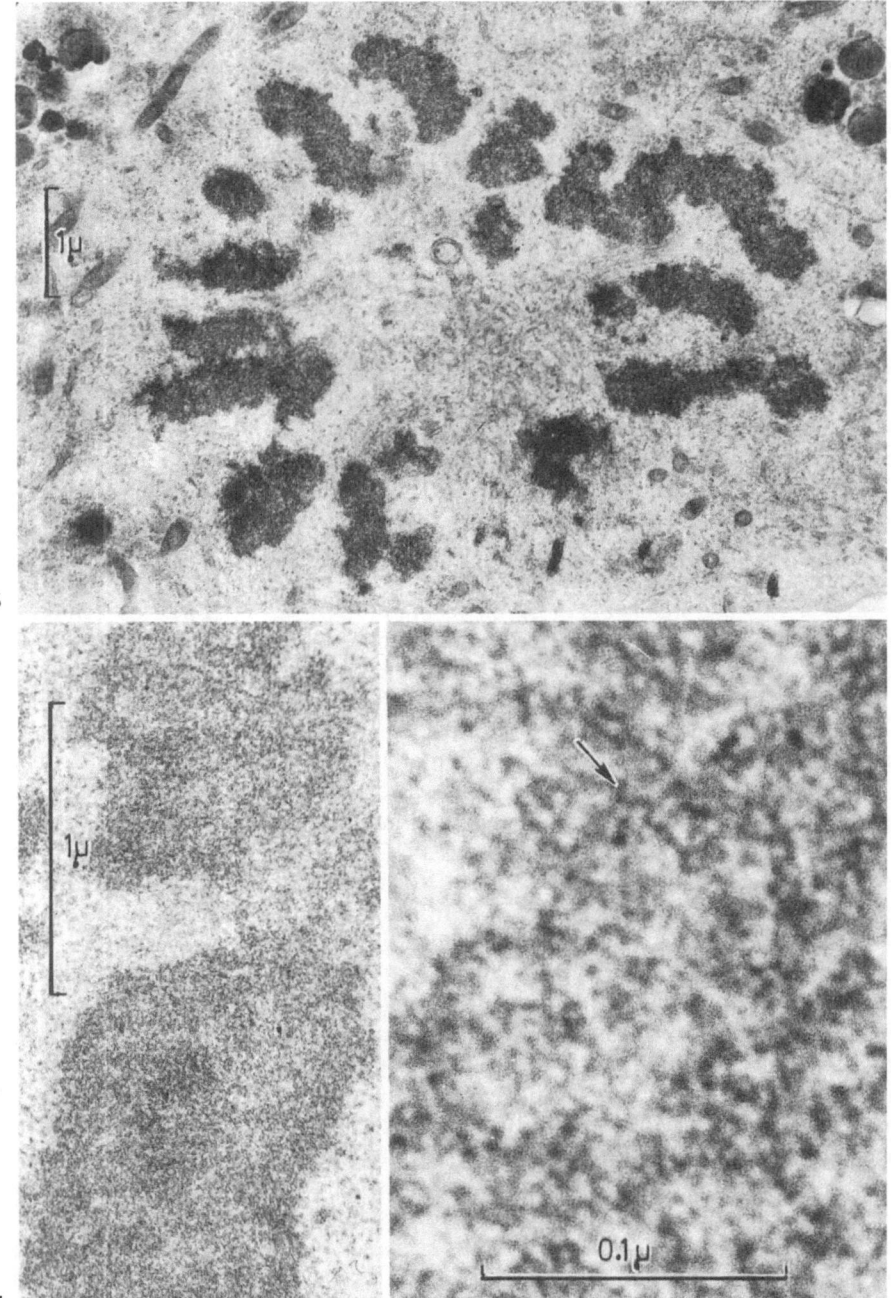

Fig. 46

Fig. 47

Fig. 48

Figs. 46–48. Electron micrographs of thin sections of human chromosomes fixed in buffered glutaraldehyde and embedded in Epon, at different magnifications. The arrow in Fig. 48 points to a fibril
of 20–30 Å diameter. (After SCHWARZACHER and SCHNEDL, 1967)

Some other information was obtained from sections of chromosomes which had been pretreated with hypotonic solutions (e.g. Hanks' solution 1 part +3 parts distilled water) and fixed with acetic acid-alcohol. In such preparations the fine fibrils are not equally well seen. But a remarkable difference exists between preparations which have been contrasted either with uranyl acetate or phosphotungstic acid (SCHWARZACHER and SCHNEDL, 1969). In sections treated with uranyl (which contrasts chiefly DNA) fine fibrils of maximal 30 Å thickness can sometimes be seen. They seem to build up thicker fibrils which are vaguely outlined and measure up to about 200 Å thickness (Fig. 52). In sections treated with tungsten, which contrasts chiefly proteins fine fibrils are not seen but the whole chromatin seems to be made up of irregularly folded fibrils of about 200–300 Å thickness (Figs. 49 and 50). These fibrils reveal sometimes a tubular structure (Fig. 51). This can be interpreted in such a way that a DNA-molecule is coiled or folded into a thicker fibril and that proteins form a sheath around

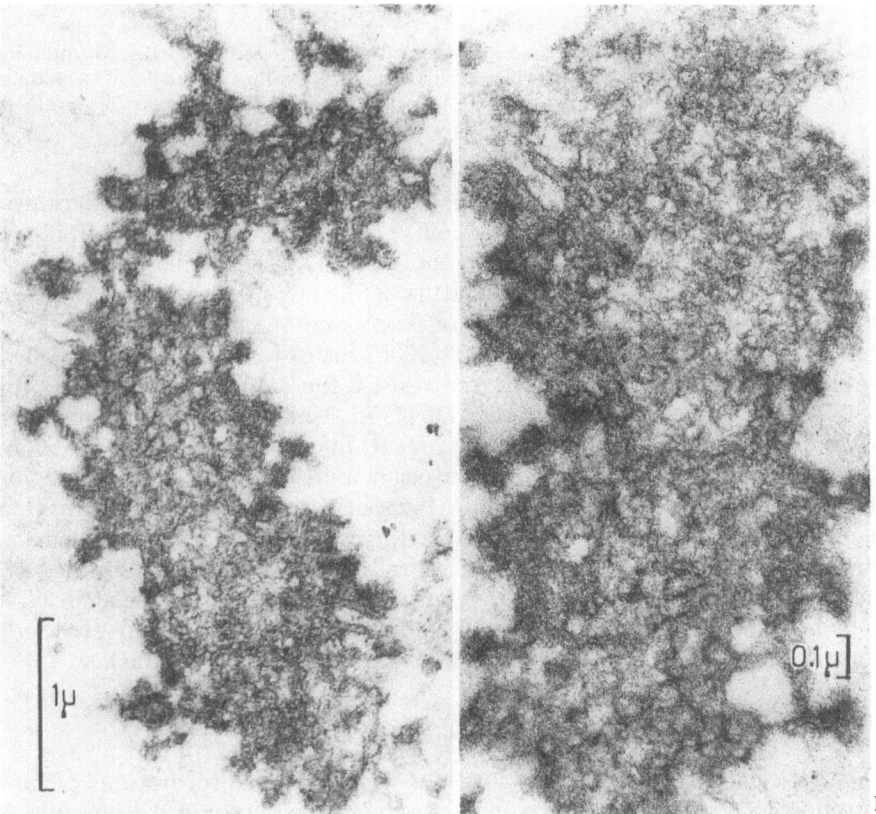

Fig. 49   Fig. 50

Figs. 49 and 50. Electron micrographs of thin sections of human chromosomes, treated with hypotone solution (Hanks: A. dest 1:3), fixed in acetic acid-ethanol (1:3) and embedded in Epon. Section contrasted with tungsten acid. Irregular running fibrils of about 150–200 Å diameter

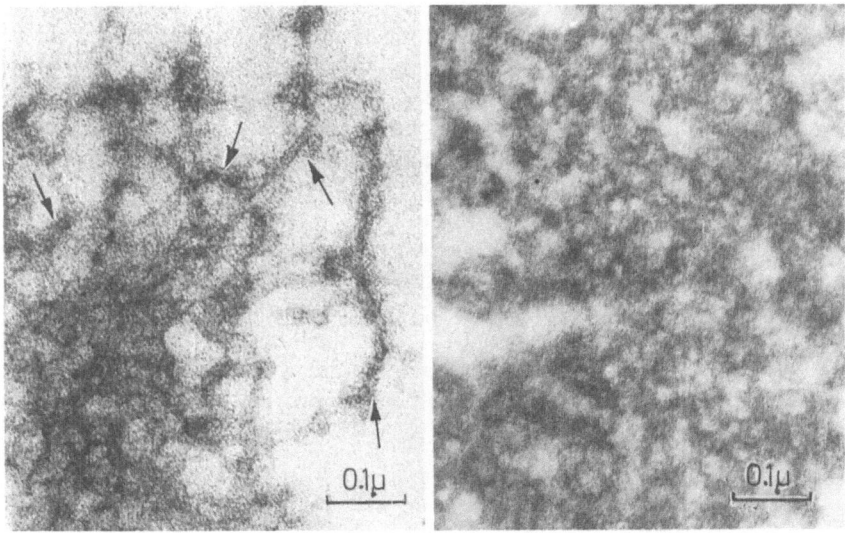

Fig. 51   Fig. 52

Figs. 51 and 52. Same techniques as in Figs. 49 and 50. In Fig. 51 section contrasted with tungsten acid. Arrows point to fibrils in cross and length sections appearing as tubules with the outside contrasted. In Fig. 52 section contrasted with uranyl acetate. The fibrils appear finer without the outside contrast. (After SCHWARZACHER and SCHNEDL, 1969)

this fibril. At least the molecules available to bind to tungsten are concentrated at the periphery of the fibril. This of course does not exclude the possibility that protein globules form also a bead-like core of the DNA-protein fibre (KORNBERG, 1974). Another interpretation would be that a fine DN-histone fibril is supercoiled in a way to appear as a hollow tube (PARDON and WILKINS, 1972). The greater thickness of the elementary fibrils (200–300 Å instead of 100–150 Å) may be due to the hypotonic treatment which may cause a loosening of the coils and folds.

Fibrillar components thicker than 150 Å. In sections of preparations directly fixed by aldehydes (without any pretreatment) thicker fibrils are only indistinctly seen. KAUFMANN *et al.* (1960) reported such units which may be formed by a coiling of the 100 Å elementary fibril. LAMPERT and LAMPERT (1970) have also suggested that a 200–300 Å fibre may be formed by a coiling of the elementary fibril. Such thicker units are not necessarily identical with the distinct fibrils of 200–300 Å diameter seen in preparations which were pretreated in various ways before fixation (e.g. the afore-mentioned hypotonically pretreated section preparations and all sorts of whole mount and spread preparations).

Whole Mount Preparations Spread on a Glass Surface

The first electron microscopic pictures of whole mount preparations of human chromosomes, prepared as for standard cytogenetic chromosome investigations were published by BARNICOT and HUXLEY (1961). A detailed study using a similar method in our laboratory (SCHWARZACHER and SCHNEDL, 1967) revealed the following: The main elements are fibrils of 200–300 Å thickness. They are laid in many short folds or in somewhat irregular coils. Sometimes it can be shown that they consist of coiled 30 Å fibrils (see Figs. 53 and 54). It may be concluded that these fibrils of 200–300 Å diameter are the elementary fibrils which measure between 80–150 Å in section preparations of directly fixed chromosomes. Due to the hypotonic pretreatment, the coiling of the fine fibril is loosened and the elementary fibril thus appears thicker than in preparations of instantaneously fixed chromosomes. The 200–300 Å fibrils are comparable with the fibrils of

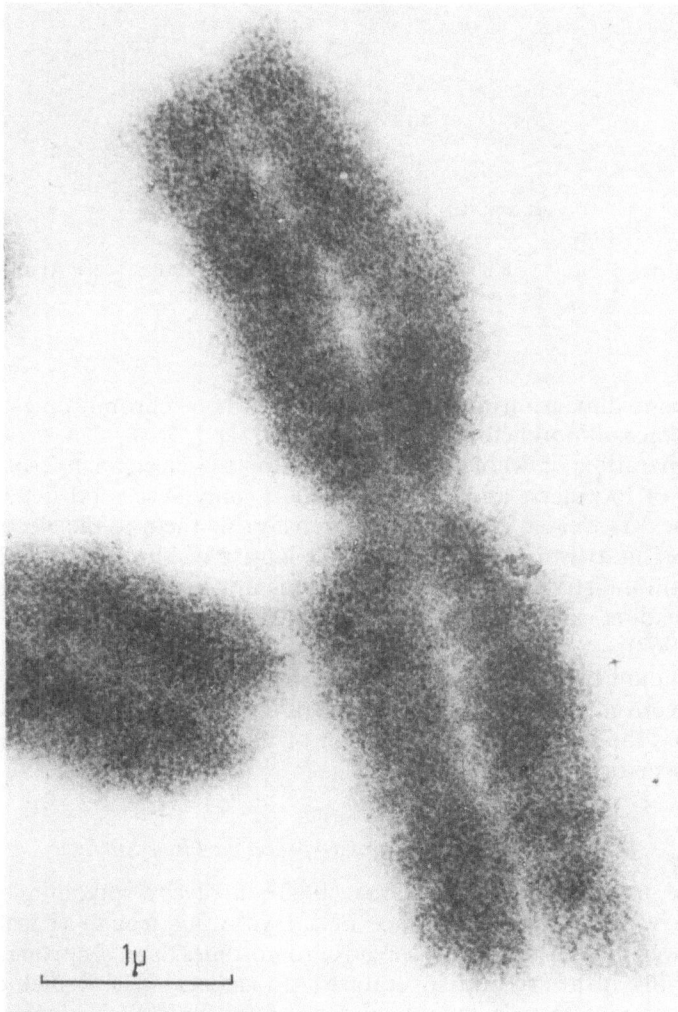

1μ

Fig. 53. Electron micrographs of a total preparation of a human chromosome No. 3. Standard chromosome preparation (hypotone pretreatment, acetic acid-ethanol fixation, air drying), hydrolysis in 1 N HCl at room temperature for 5 min, coating with collodium and peeling off. (According to SCHWARZACHER and SCHNEDL, 1967)

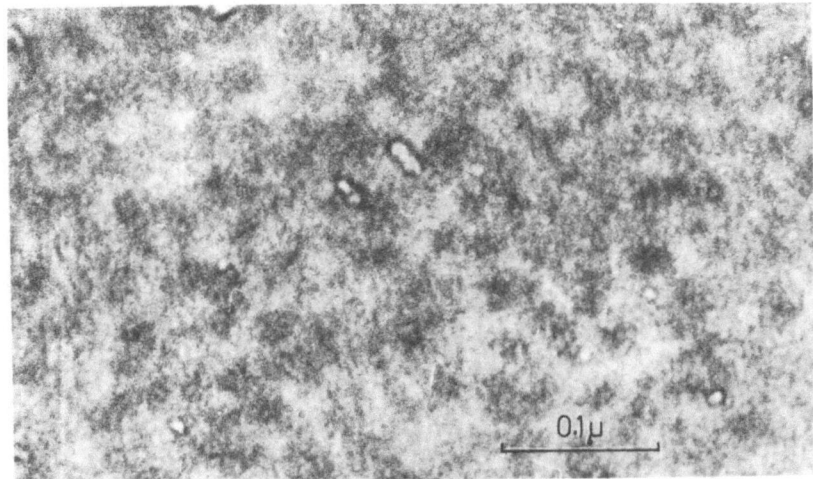

Fig. 54. Detail from Fig. 53 at high magnification. The thick fibrils ($\sim$200 Å) are built up of irregular screws of $\sim$30 Å fine fibrils

about the same dimensions in section preparations of chromosomes pretreated with hypotonic solution before fixation (Fig. 50).

The preparation of whole chromosomes for the electron microscope after the method of BARNICOT and HUXLEY (1961) is only successful if the prepared chromosomes are treated by HCl before transferring them to the electron microscope grid. This treatment enhances the contrast of the fibrils, although the fibrils are still indistinctly seen in preparations in which the chromosomes were directly spread on carbon film without any HCl-treatment (SCHWARZACHER and SCHNEDL, 1967).

Whole mount preparations of chromosomes were also investigated with the scanning electron microscope (e.g. CHRISTENHUSS et al., 1967; PAWLOWITZKI et al., 1968). Due to the limited resolution of the scanning electron microscope no new information could be obtained.

Whole Mount Preparations Spread on Fluid Surface

This method, used the first time by GALL (1963), involves a spreading of chromosomes on a water or alcohol surface in a LANGMUIR trough (KLEINSCHMIDT, 1959), followed by transferring the spread chromosomes to an electron microscope grid and drying at the critical temperature (ANDERSON, 1951). It should be noted that this treatment may include a contact of the unfixed chromosomes with distilled water, thus having a very strong hypotonic effect. The spreading of the chromosomes is also quite extensive. Of all methods used this is certainly the one which produces the *most artefacts* and the results obtained with it have to be interpreted very critically. Nevertheless, since single fibrils can be isolated by this method its value is indisputed.

Most investigators using this method reported that chromosomes of many different species consist of fibrils of about 200–300 Å thickness (e.g. GALL, 1963, 1966; WOLFE, 1965; OSGOOD et al., 1964; DU PRAW, 1965a and b; LAMPERT, 1969; ABUELO and MOORE, 1969; COMINGS and OKADA, 1970; BAHR et al., 1973; BAHR and GOLOMB, 1974; JAFFRAY and GENEIX, 1974). Figs. 55, 56, and 57 show human chromosomes prepared according to this method. The 200–300 Å-fibril corresponds with all probability to the 100 Å elementary fibril seen in directly fixed and sectioned chromosomes. The greater thickness is presumably due to a swelling of the fibril or a partial uncoiling of the presumed fine fibril (of 30 Å). Thus, the same conditions exist as in section preparations of hypotonically pretreated chromosomes or as in the hypotonically pretreated whole mount chromosome preparations. More evidence, that the fibril corresponds to the 100 Å fibril of directly fixed section preparations comes from the study by WOLFE (1968) who showed that the fibril diameter is between 100 and 150 Å when the chromosomes were first fixed in formaldehyde and then spread on the Langmuir trough and dried. Only RIS (1966, 1969) suggested another possibility. He proposes that two 100 Å fibrils may build up one 200–250 Å fibre. His pictures can be interpreted in this way, but not necessarily so. It is, for instance, quite possible that in his preparations differently stretched fibrils came to lie together by chance during the course of preparation.

The fine fibril of about 30 Å is seen only under certain conditions in water surface spread preparations. LAMPERT and LAMPERT (1970) reported them in sites where the elementary fibril is particularly stretched in the preparation. It seems that the coiled fine 30 Å-fibril is drawn out in that case. WOLFE (1965) found that the fine fibrils are destroyed by DNAase whereas the 250 Å-fibril is more resistant. EDTA causes a decrease in diameter of the 250 Å-fibril, which may be caused by a destruction of its peripheral parts (SOLARI, 1968).

The observations on water surface spread and critical point dried chromosome preparations seem to be comparable to those obtained by the other methods, if the hypotonic effect is taken into consideration.

Methods Using Special Treatments

A number of observations have recently been published on chromosomes which were treated in a similar way as for the different banding techniques, including treatments with NaOH, concentrated salt solutions at various temperatures and proteolytic enzymes. Most of them were undertaken to elucidate the mechanism of the banding technique and have been already discussed in chapter V (e.g. CERVENKA et al., 1973; ROSS and GORMLEY, 1973; COMINGS et al., 1973; BURKHOLDER et al., 1973; RUZICKA, 1974; RUZICKA and SCHWARZACHER, 1974). The preparations made from chromosomes treated with proteolytic enzymes have not given any clues regarding the fine structure of the fibrils. The methods, however, using NaOH and heat gave surprisingly uniform pictures, revealing that chromosomes appear to be built up of fibrils 300–600 Å thick (RUZICKA, 1974; RUZICKA and SCHWARZACHER, 1974).

RUZICKA (1974) found that the fibrils become distinctly seen only after Giemsa staining, following the banding treatment. RUZICKA and SCHWARZACHER (1974)

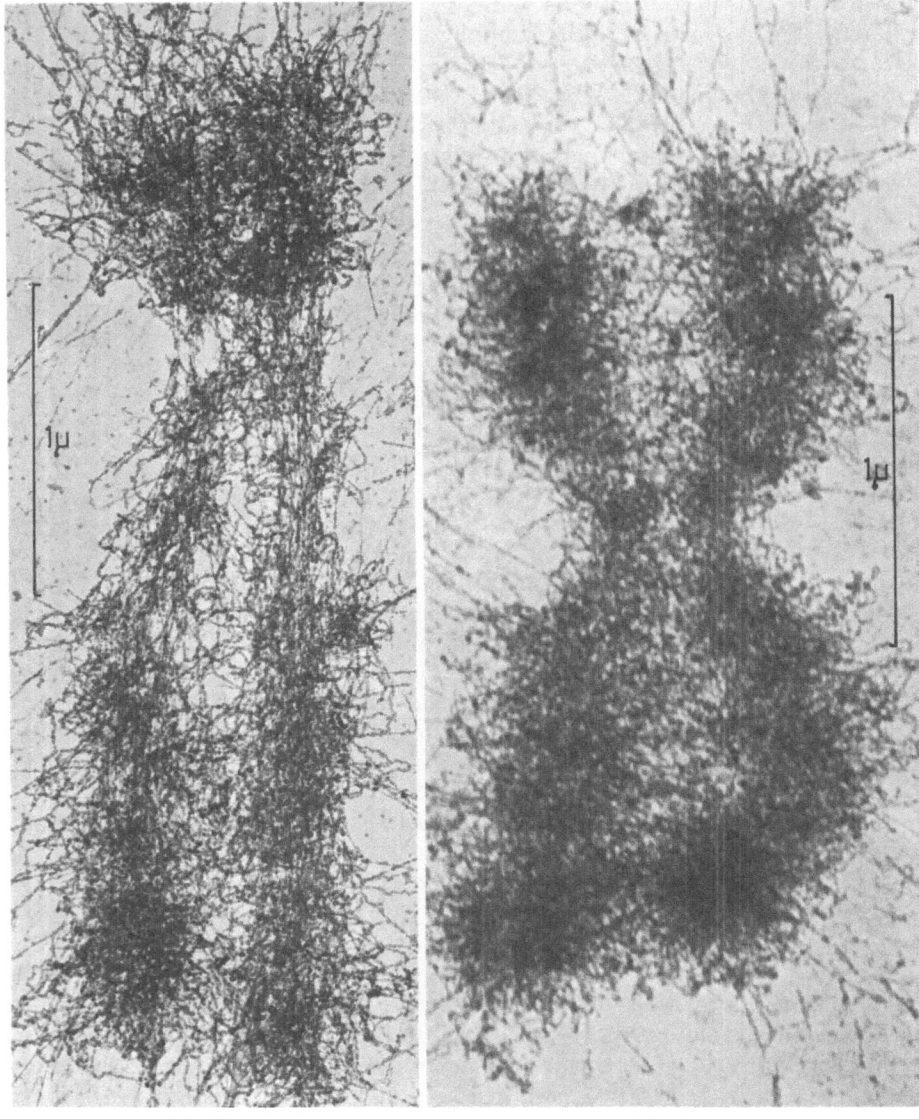

Fig. 55   Fig. 56

Figs. 55 and 56. Electron micrographs of total preparations of human chromosomes, spread on dest. water in a Langmuir trough and dried at the critical point temperature. Fibrils of 200–300 Å diameter are seen folded to irregular loops, many sticking out from the chromatids. Courtesy of
G. Bahr

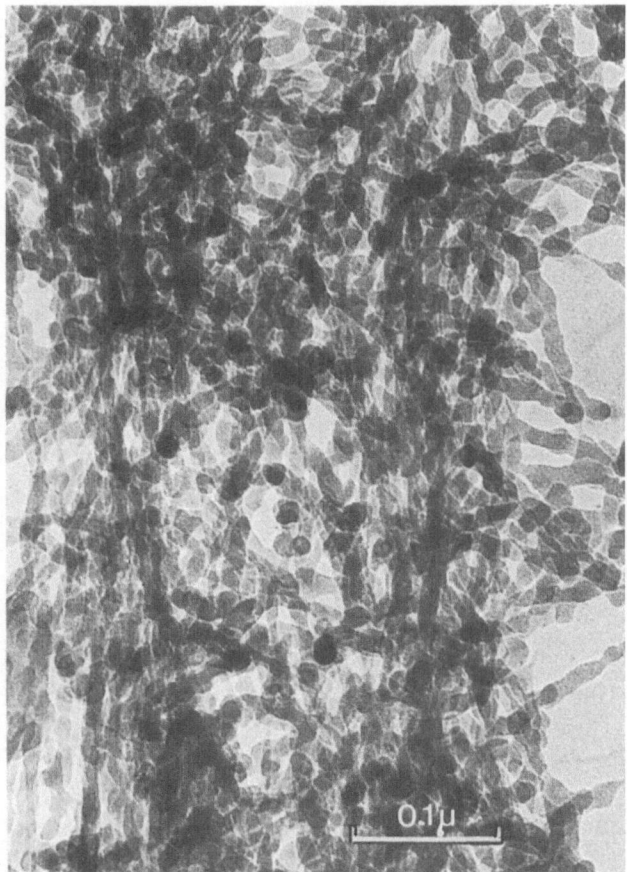

Fig. 57. High magnification of a chromosome similarly prepared as in Figs. 55 and 56. Courtesy of G. BAHR

discuss the possibility that the bromine component of the Giemsa dye is responsible for the enhancement of the contrast as well as the irregular thickening of the fibrils. In Figs. 58 and 59 human chromosomes, treated by the banding technique of SCHNEDL (1971) and Giemsa stained are shown. Such preparations are pretreated with hypotonic solution and fixed in acetic acid-alcohol. Without the treatment of the banding technique the diameter of the elementary fibrils would be 200–300 Å. Thus the special treatment may cause an additional swelling and the Giemsa dye may be aggregated onto the surface of the fibrils. The additional swelling may have this reason: As McKAY (1973) suggested, the banding techniques act by a removal of cations causing a destruction of the chromatin structure followed by a reconstruction of the structure by addition of divalent cations. One might speculate that the reconstitution is not complete, in other

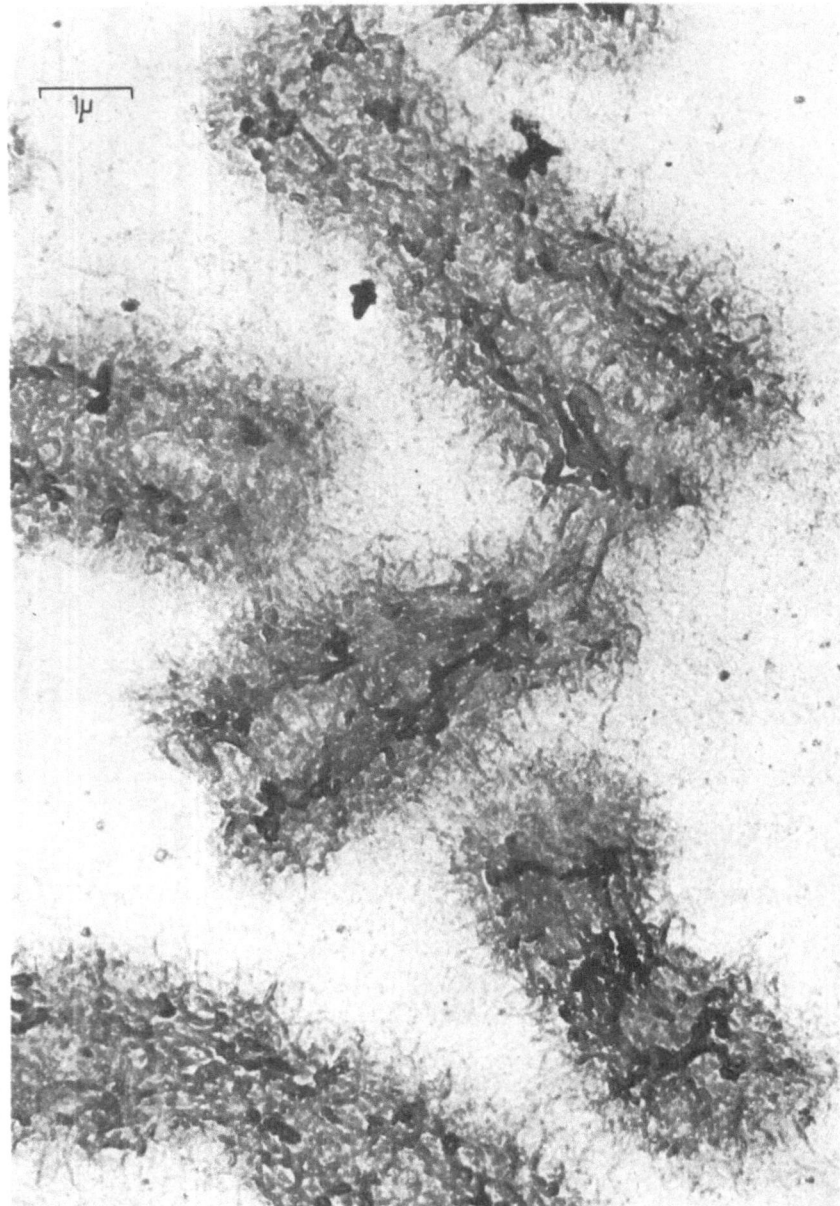

Fig. 58. Electron micrograph of a total preparation of metaphase chromosomes, spread and air dried on slides, G-banding stained, carbon coated and peeled off the slide. Thick fibrils irregularely running through the chromatids. Loops of fibrils are sticking out of chromatids. (From RUZICKA and SCHWARZACHER, 1974)

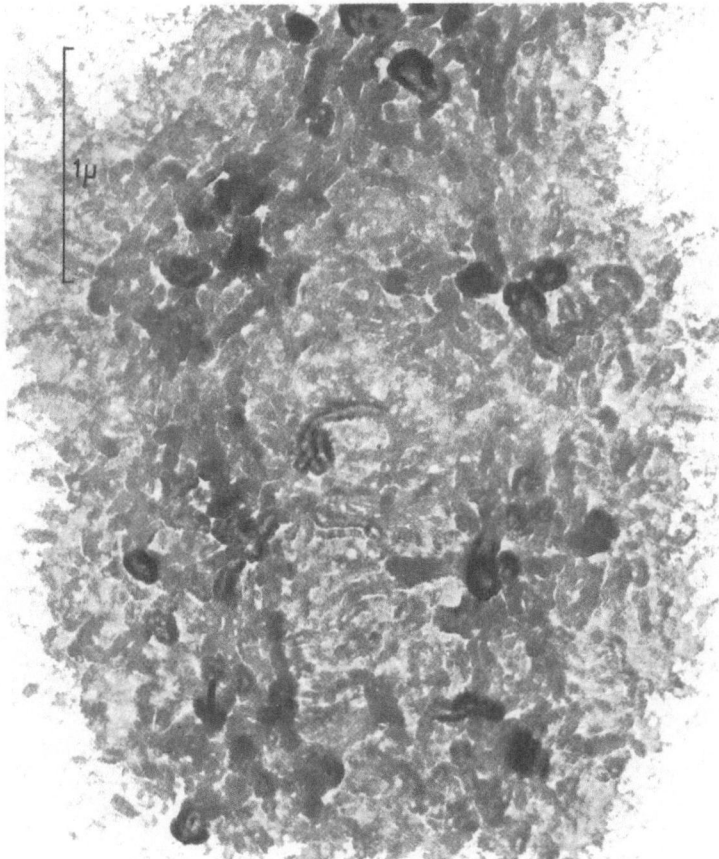

Fig. 59. Same technique as in Fig. 58. In the G-band regions the chromatids are thicker and the fibrils appear to be stronger contrasted. (From RUZICKA and SCHWARZACHER, 1974)

words, that the fine fibril can recoil only to a certain degree thus forming a rather thick elementary fibril. The agglomeration of the Giemsa dye on the outside of the fibril is indicated by a double structure (RUZICKA and SCHWAR-ZACHER, 1974, see also Fig. 59). If these speculations are accepted, the 300–600 Å fibril is in fact the elementary fibril of 100 Å.

High magnifications in the ordinary transmission electron microscope as well as with a special perspective representation of the whole mount preparation by tilting giving the impression of three-dimensional pictures (see Fig. 60) do not give much information about the fine structure of the thick elementary fibril. Sometimes several parallel lying thinner fibrils are indicated, sometimes a rather honey-combed appearance is observed perhaps caused by a 30 Å-fibril coiled into the thick fibril.

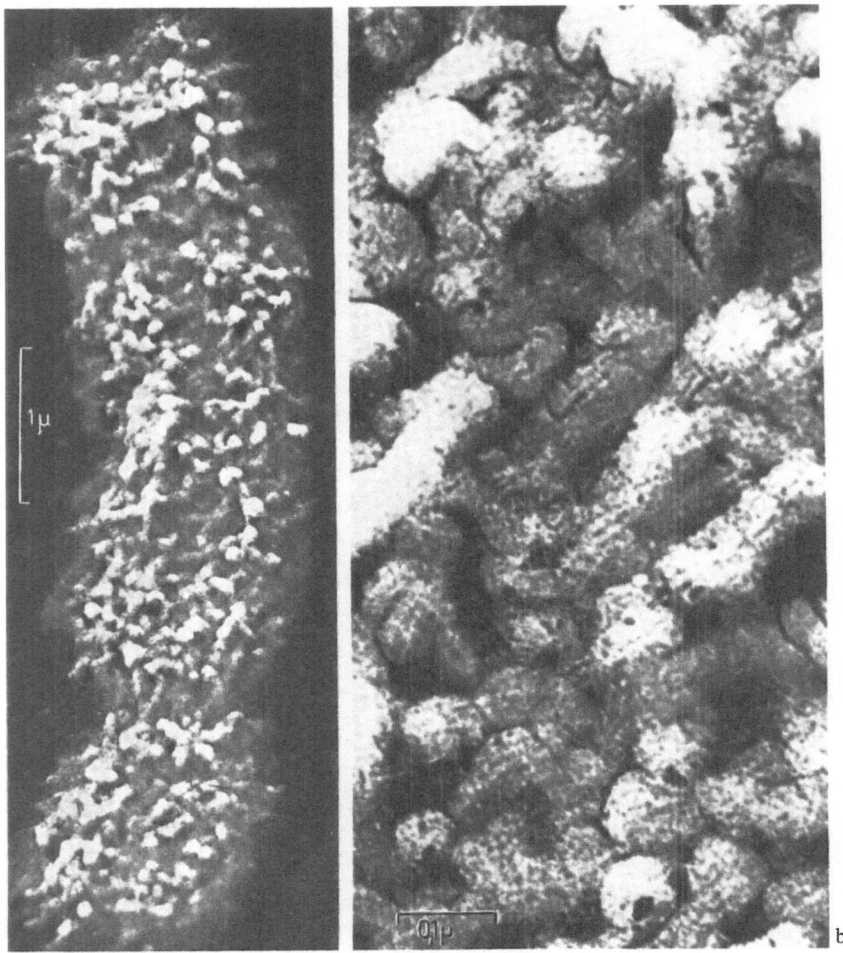

Fig. 60a and b. Pseudo three-dimensional representation (perspective view by tilting the specimen) of (a) human chromosome, at (b) at high magnification ($\times 125\,000$). The thick fibrils seem to be composed of finer fibrils.

Another approach to the study of chromosome fibrils by special methods was made by SORSA an coworkers (see SORSA *et al.*, 1970; SORSA, 1972 and 1973 b). They used *Drosophila* giant chromosomes spread on water surface and dried at the critical temperature and afterwards treated with NaOH and urea. They reported fine fibrils of about 20 Å diameter (possible bare DNA molecules) and of 30 Å (possible DNA-histone fibrils). The next order they found were 100 Å thick, sometimes stretched to 70 Å. It must be emphasized that the pictures of SORSA show that the chromosomes were stretched to a very high degree. Hence the small diameter of the elementary fibril could be explained in that way.

X-Ray Diffraction Studies

X-ray diffraction studies have been done on isolated chromatin fibres from eukaryotic interphase nuclei and on experimentally produced DNA-histone (DNH) fibrils. RICHARDS and PARDON (1970) and PARDON and WILKINS (1972) described as the most probable model a supercoiled DNA-fibril with a diameter of 100–130 Å and a pitch of 120 Å. BRAM and RIS (1971) reported rather an irregularly supercoiling with an average pitch of 45 Å of a 100 Å thick fibril as most probable. It should be noted that PARDON and WILKINS (1972) discuss the possibility that the supercoiled DNH fibril may appear in the electron microscope as a hollow cylinder. This has indeed been found under certain circumstances by DAVIES (1968) and ourselves (SCHWARZACHER and SCHNEDL, 1969; see Fig. 51).

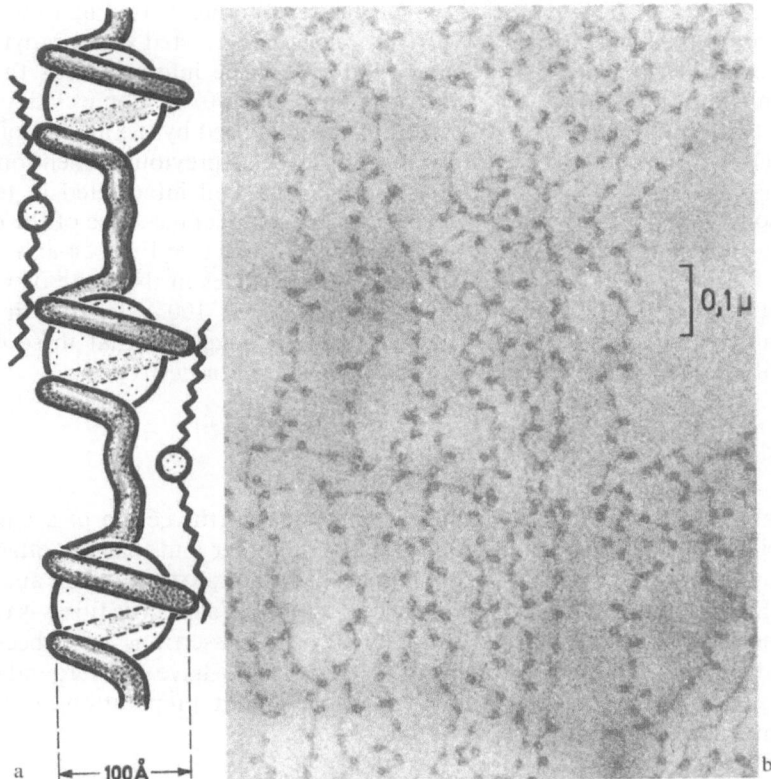

Fig. 61. (a) Schematic representation of a possible model for the subunit structure of the chromatin fibre. The globules forming the core are complexes of histones H2A, H2B, H3 and H4 with their basic segments complexed with DNA (the dark band). Possibly crosslinking histone Hl-complexes are drawn on the outside (after BALDWIN et al., 1975). (b) Chromatin fibres isolated from a chicken erythrocyte nucleus. The v-bodies are ∼ 70 Å in diameter consisting of loops of a fine fibril (presumably the DNA molecule). The connecting strand is ∼ 140 Å in length. Negative staining with uranyl acetate. Courtesy of D.E. OLINS and A.L. OLINS

Repeated Substructures of Chromatin Fibres (ν-Bodies)

PARDON *et al.* (1967) reported that isolated chromatin fibrils contain a repeating substructure. KORNBERG (1974) proposed a model of the chromatin fibre consisting of centrally situated histone globules (fractions IIA, IIB, III and IV) around which the DNA molecule winds in a few wide coils. Between the globules the DNA molecule is free of a central histone. On the outside histone (fraction I) may hold the globular subunits together. The DNA of one subunit comprises about 200 base pairs. BALDWIN *et al.* (1975) suggested a similar model with globular subunits of about 100 Å in diameter and a somewhat greater distance (see Fig. 61 a).

Some direct morphological evidence has been reported for repeated thickenings along the chromatin fibril. NOLL (1974) found them in nuclease treated rat liver nuclear chromatin. OLINS and OLINS (1974) reported a linear array of small chromatin particles of 70 Å diameter in isolated fibres of water swollen eukaryotic nuclei. They called these particles "ν-bodies". In Fig. 61 b isolated chromatin fibres of a chicken erythrocyte nucleus contrasted with uranyl acetate are shown (OLINS and OLINS). The fine fibril can be interpreted as the DNA molecule forming wide loops (with a diameter of 70 Å), connected by more straight stretches (in average 140 Å long), as is proposed by the models of KORNBERG (1974) or BALDWIN *et al.* (1975). As has been previously mentioned, the pictures of thin sections of chromosomes can be well interpreted in terms of this model (see Fig. 48). Also the granular or bead-like appearence of the elementary chromatin fibril in different sorts of preparations (see Figs. 54 and 57) may speak for the existence of repeated chromatin particles in the sense of ν-bodies. GRIFFITH (1975) found similar segments (of about 100 Å length) in "minichromosomes" of viral DNA. Enzymatic studies suggested that these particles are only present when the protein components are not destroyed.

Core Fibres

Several authors have discussed the possibility that in the centre of a chromatid a special fibre runs through from one end to the other and that the other fibrils form loops around this core (ABUELO and MOORE, 1969; STUBBLEFIELD and WRAY, 1970; SORSA *et al.*, 1970; SORSA, 1973). The thickness of such core fibres is reported differently. It must be pointed out that all these assertions have been made on water surface spread chromosomes. Core fibrils have been found neither in directly fixed chromosomes, nor in whole mount preparations other than the water spread method.

Conclusion about the Structure of Chromosome Fibrils

As has been emphasized several times, the observations made on preparations of directly *in situ* fixed chromosomes are certainly the most reliable. It must be explained why this point, which is self-evident to most electron microscopists, is repeated here again: Geneticists and other scientists who want to interpret electron microscopic pictures without knowing the pitfalls of the methods, have a tendency to turn to the pictures of water surface spread whole mount preparations, because they are fascinated by the well aligned and clear looking fibrils. This method, as has been said above, causes a lot of artefacts because of the hypotonic treatment and the stretching of the fibrils.

The smallest unit seen so far is a fibril of about 20–30 Å diameter. It represents presumably the DNA molecule (the double helix). One such fibril is coiled or perhaps more irregularly folded to form an elementary fibril of about 100 Å (80–150 Å) thickness. This fibril is characterized by having periodical thickenings caused by centrally located globules of histones arount which the DNA molecule winds in a few wide coils. Proteins are situated presumably also on the outside of the elementary fibril since it may appear as a hollow tubule after staining with tungsten.

The fine 20–30 Å-fibril can be loosened or partly uncoiled by hypotonic solutions thus forming a somewhat thicker elementary fibril of about 200–300 Å. Other treatments such as those used for the Giemsa banding techniques cause a still stronger thickening of the elementary fibril, up to 300–600 Å thickness. This thickening may again be caused by a still looser packing of the 20–30 Å-fibril. Differences in the thickness of the elementary fibrils can also be due to a differential stretching (as caused by the method of spreading on a water surface).

This shows that the 20–30 Å-fibril is presumably rather stable and that the differences observed in the diameter of the elementary fibril are caused by a differential coiling or folding of the fine-fibril due to the different methods applied.

The 100 Å elementary fibril has been reported to form still thicker fibrils by a secondary coiling in some cases. Since such supercoiling is not seen in whole mount preparations it does not seem to be a constant or stable feature.

3. Arrangement of Fibrils

The arrangement of the elementary fibrils within the chromosome, or better the chromatid, is of great interest. In the foregoing chapter it has been stated that the elementary fibril (80–150 Å thick) is most probably built up of one 20–30 Å fibril, which corresponds to one DNA double helix molecule. Therefore the number of elementary fibrils per chromatid should correspond to the number of DNA molecules.

Survey of Electron Microscopic Findings

From ultrathin sections of directly fixed chromosomes one can deduce that the elementary fibrils are laid in very short foldings or coils (BARNICOT and HUXLEY, 1965; SCHWARZACHER and SCHNEDL, 1967; see Figs. 47 and 48). However, a regular coiling cannot be observed. SPARVOLI et al. (1965) attempted to reconstruct the arrangement of the fibrils in chromosomes of *Tradescantia* from serial sections. They reported that one fibril forms a rather thick strand and that a chromatid may contain several such strands. Besides the circumstance that serial sections could be analyzed only within limited chromosomal regions in this study it should be remembered that mitotic chromosomes of *Tradescantia* are possibly polytene. Thus one strand of a *Tradescantia* chromosome (called "chromonema" by SPAR-VOLI *et al.*, 1965) would correspond to a chromatid of a unitene chromosome.

Sections of hypotonically pretreated chromosomes (SCHWARZACHER and SCHNEDL, 1969) show the elementary fibrils to be rather loose (Figs. 49 and 50). The fibrils are irregularly folded and no preferential direction can be observed in their arrangement. It is of course impossible from these pictures to decide whether one or more fibrils constitute a chromatid.

In whole mount preparations of chromosomes spread on a glass surface (BARNICOT and HUXLEY, 1961; STEVENS, 1967; SCHWARZACHER and SCHNEDL, 1967) nothing more can be learned than that the elementary fibrils are bent and folded in very short loops. No regular coiling and no preferential direction of the fibrils can be observed (Figs. 53 and 54).

The whole mount preparations of chromosomes spread on a water surface and critical point dried (e.g. GALL, 1963; DU PRAW, 1965; OSGOOD *et al.*, 1964; LAMPERT and LAMPERT, 1969; BAHR *et al.*, 1973) show the fibrils usually less folded. They run rather in the longitudinal direction and often several fibrils appear to run parallel (Fig. 55). A free ending of a fibril can never be observed.

Finally, whole mount preparations of chromosomes treated as for the Giemsa banding techniques (RUZICKA and SCHWARZACHER, 1974) again show a very irregular folding of the elementary fibrils (Figs. 58 and 59). A preferential direction of the fibrils is sometimes indicated, particularly in the centromeric region and in interband zones. There the fibrils appear to run somewhat parallel in the longitudinal direction. Free endings of fibrils are never seen. What appear as sticking out fibrils at low magnifications are in reality loops of fibrils.

4. Single Stranded or Multistranded Chromatids?

Electron Microscopic Observations on Metaphase Chromosomes

From the examination of electron microscopic pictures of chromosomes prepared after different methods it is not possible to decide whether one single elementary fibril builds up a chromatid or whether more than one fibril is present. The section preparations as well as the whole mount preparations after standard or Giemsa banding treatment can be interpreted in either way. That free endings of fibrils cannot be found within chromatids favours single strandedness.

The pictures obtained after spreading chromosomes on a water surface followed by critical point drying sometimes revealing several fibrils running parallel in the length direction of the chromosome (Fig. 55) have often been interpreted in the sense of a multistrandedness (e.g. RIS, 1961; OSGOOD *et al.*, 1964; STUBBLE-FIELD, 1973). DU PRAW (1965) and COMINGS (1972) pointed out that these pictures can also be interpreted as loops of one fibril, the loops being stretched in the

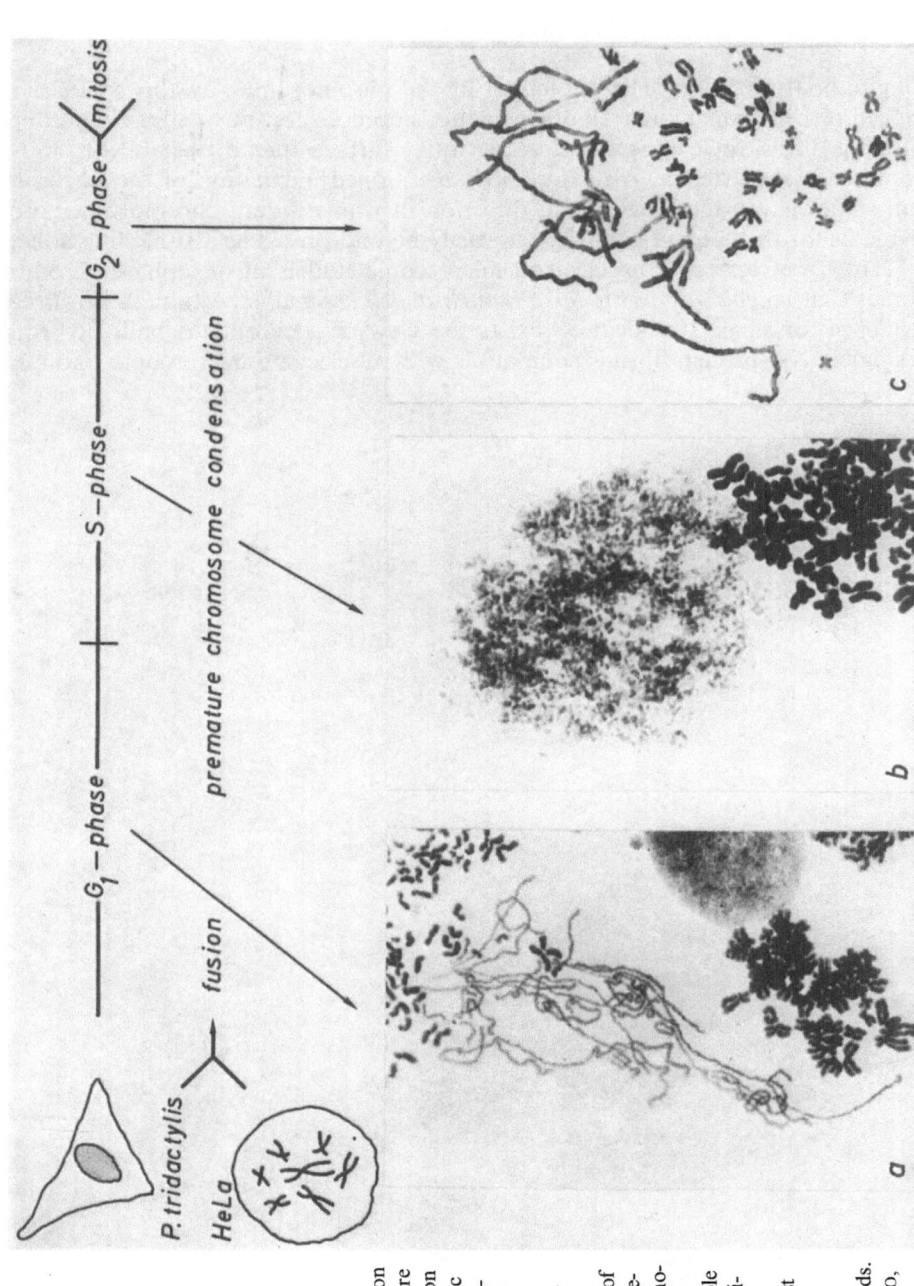

Fig. 62a–c. Representation of the course of premature chromosome condensation after fusion of interphasic cells (in this case rat kangaroo cells) with mitotic cells (in this case HeLa cells). (a) premature condensed chromosomes in G₁-phase are composed of single chromatids. (b) premature condensed chromosomes in S-phase appear "pulverized", the invisible stretches between the visible chromatid segments being the presumably just synthetizing segments. (c) G₂-chromosomes are built up of two chromatids. (From SPERLING and RAO, 1974)

longitudinal direction. The stretching of the chromosomes by this method is indeed rather strong. Since all other methods do not show any definitely parallel running fibrils, their appearance in the water surface spread preparations must be taken as an artefact. Also the above mentioned inclination of the fibrils to run somewhat in the longitudinal direction in whole mount chromosomes prepared as for the Giemsa banding can easily be interpreted as a stretching effect.

Thus, we can say from electron microscopic studies of metaphase chromosomes that *no evidence exists for a multistranded chromatid*. Although no direct evidence for single strandedness exists the view of a chromatid built up from a single DNA-protein fibril is compatible with all electron microscopic findings.

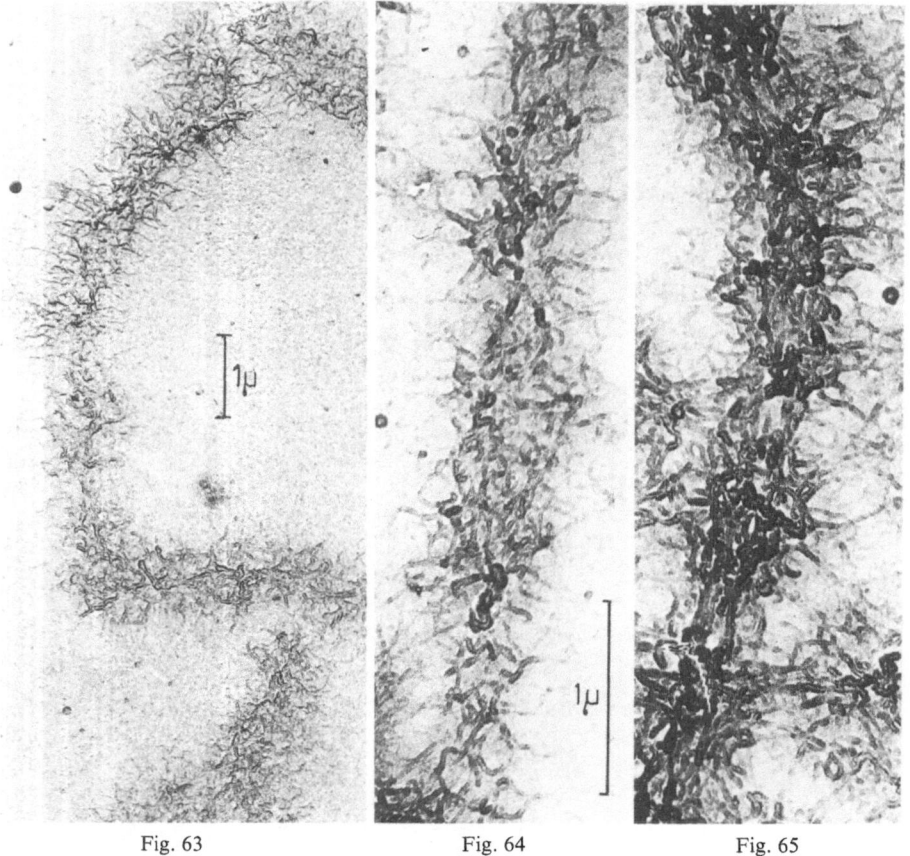

Fig. 63   Fig. 64   Fig. 65

Figs. 63–65. Electron micrographs of premature condensed chromosomes. Technique as in Figs. 58 and 59. The G-band regions characterized by a tighter folding and a stronger contrast of the fibril. Chromosomes of Figs. 63 and 64 are in G1, of Fig. 65 in G2

Observations on Prematurely Condensed Chromosomes

It has been shown by JOHNSON and RAO (1970) that fusion of a cell in metaphase with a cell in interphase causes the extended chromosomes of the latter to contract. (The fusion of cells can be achieved with the aid of Sendai-virus after the method introduced by HARRIS *et al.*, 1966.)

Detailed analysis of this phenomenon, called "premature chromosome condensation" or "prophasing" has shown that if chromosomes which are in the G1-period of the cell cycle (that is before DNA replication) are forced to condense, they appear in the light microscope as single threads (Fig. 62), whereas prematurely condensed chromosomes in the G2-period (that is after DNA replication) appear as double threads (Fig. 61). Precociously condensed chromosomes in the period of DNA-synthesis (S-period) appear fragmented in the light microscope (Fig. 61), the fragments representing condensed and not synthetizing areas, the regions between the fragments are thought to consist of uncoiled DNA fibrils which are in a state of replication (SANDBERG *et al.*, 1970; MATSUI *et al.*, 1972; RAO and JOHNSON, 1974; SPERLING and RAO, 1974).

Electron microscopic studies done so far on whole mount preparations of prematurely condensed G1 and G2-chromosomes, prepared after the standard method and then treated as for the Giemsa banding methods (SCHWARZACHER *et al.*, 1974) revealed the following: As has been explained above, the Giemsa banding treatment shows the thickened elementary fibrils particularly clearly. In Figs. 63–65 very thin regions of prematurely condensed G1 chromosomes are shown. The chromatid consists of irregular loops of elementary fibrils. The pictures are suggestive of only one fibril being present in one chromatid laid in many loops and returning to a main axis in-between. The comparison with lampbrush chromosomes (e.g. CALLAN, 1963) is forced upon the observer. A special core fibril is not seen, but the elementary fibril itself forms a sort of backbone in the chromatid. However, the pictures obtained so far cannot completely rule out the possibility of more than one elementary fibril composing a chromatid. In Fig. 71 a model of a prematurely condensed chromatid consisting of only one fibril is proposed.

Lack of Evidence for Half Chromatids in Human Chromosomes

Subunits of chromatids, sometimes called "chromonemata" or "half chromatids" or even "quarter chromatids" have been reported in some instances. For example in chromosomes of *Vicia faba* several strands seem to run parallel as has been suggested from electron microscopic studies by KAUFMANN *et al.* (1960) and light microscopic studies by TROSKO and WOLFE (1965). Light microscopic observations on chromosomes of *Tradescantia* (NEBEL, 1958), of *Hyacinthus* (SORSA and SORSA, 1968) and particularly of endosperms of *Haemanthus* (BAJER, 1965) leave no doubt that the chromatids appear divided in two or even more strands. It must be taken in consideration that in these cases the chromosomes may be better called polytene. In this connection it should be remembered that ZWEIDLER (1964) found two semiconservative replicating subunits per chromatid in *Allium cepa*. The only report of possible subchromatids in a mammalian species is that of DEAVEN and STUBBLEFIELD (1969) who concluded their presence from DNA replication studies on Chinese hamster chromosomes. Other reports (e.g. TAKAYAMA, 1973) are obviously based on a false interpretation of the microscopic picture.

As COMINGS and OKADA (1970) point out, whenever rod-like elements are twisted the impression of double strandedness can be given. They investigated nine different species of vertebrates and found no evidence for chromatid subunits. Reports of half chromatids in human chromosomes (e.g. KOULISCHER, 1963) seem also to be based on insufficient light microscopic pictures. Careful examination of light microscopic preparations of human chromosomes never results in the finding of more than one strand forming a chromatid. The same is true for all electron microscopic studies (c.f.) on human chromosomes. Also the prematurely condensed chromosomes in the G1-period consist of only one thin thread (Fig. 62). As has already been pointed out it is quite probable that they consist of even only one elementary fibril.

Semiconservative Segregation of Chromatids

TAYLOR and his coworkers showed for the first time that chromatids replicate and segregate in the same manner as the DNA double helix (TAYLOR et al., 1957; TAYLOR, 1963 and 1966; see Fig. 66). That means that chromatids consist of two chains. The semiconservative segregation of chromosomes has since been shown for many species including man (PRESCOTT and BENDER, 1963). Fig. 67 shows the chromosomes of a Chinese hamster cell, which was given BrdU in the cell cycle before the one ending with the mitosis. In each chromosome only one chromatid contains BrdU, noticeable by the weak Giemsa staining. In some places sister chromatid exchanges can be seen.

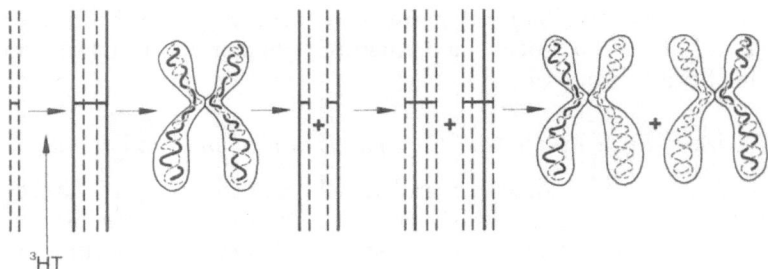

Fig. 66. Schematic representation of the distribution of subunits of DNA during chromosome duplication according to a semiconservative mode of replication. Broken lines represent unlabelled subunits, unbroken lines represent labelled subunits. (According to TAYLOR et al., 1957)

The semiconservative segregation can be particularly impressively demonstrated in endoreduplicated chromosomes (WALEN, 1966; SCHWARZACHER and SCHNEDL, 1966; HERREROS and GIANELLI, 1967). Endoreduplication consists of two independent DNA replications following each other without mitosis between. The result is that each chromosome consists of 4 chromatids (Fig. 68). If the second DNA replication takes place in the presence of ^{3}H-thymidine all chromatids will be labelled, since each one contains one newly synthetized DNA chain. If the first DNA replication is labelled then the label will appear only in two of the four chromatids. In Fig. 69 it is seen that the labelled chromatids are situated always in an outside position. This result is in accordance with a single stranded chromosome in which the daughter strands segregate semiconservatively, and are consistently placed to the outside of the parental strands.

TAYLOR (1963) also observed sister chromatid exchanges (see Fig. 67). In endoreduplicated cells the frequencies of the different types of exchanges possible are in absolute accordance with the theoretically calculated rates (HERREROS and GIANNELLI, 1967).

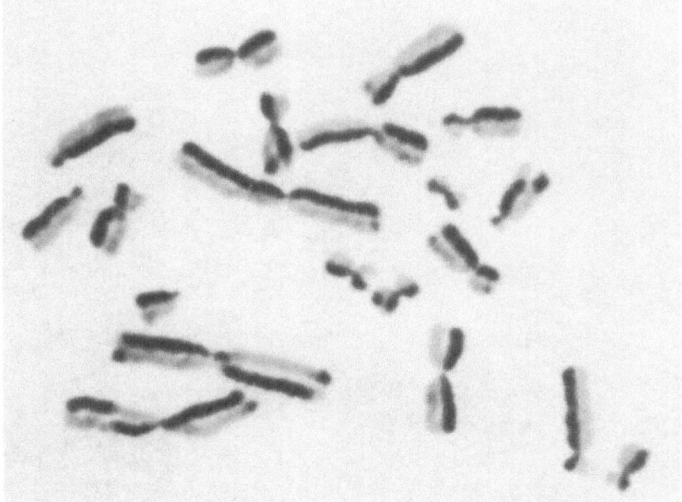

Fig. 67. Metaphase of a Chinese hamster cell (CHO-line) which was labelled with BrdU during the first cell cycle and not labelled during the subsequent second cycle. The labelled chromatids appear lightly stained. Staining with benzimidazol (Hoechst 33258) and Giemsa. Several chromatid exchanges are visible. (According to PERRY and WOLFF 1974). Courtesy of S. WOLFF

A further beautiful evidence for a semiconservative replication comes from the study of LATT et al. (1974) demonstrating the BrdU-effect only in one chromatid of the Y-chromosome after BrdU-uptake during the directly preceding S-period.

Calculation of the packing Ratio of DNA

DU PRAW and BAHR (1969), BAHR (1970) and GOLOMB and BAHR (1974) attempted to calculate the necessary packing ratio of the DNA of one genome into the chromosome set. If the human chromosome No. 13 is taken as an example, a DNA molecule of 32000 μ length has to be packed into a chromatid 5.8 μ

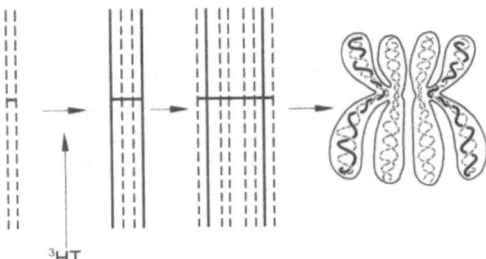

Fig. 68. Schematic representation of the distribution of subunits of DNA during endoreduplication. Labelling only during the first replication cycle. Broken lines represent unlabelled subunits. Position of labelled and unlabelled subunits within chromatids as observed in the preparations. (According to SCHWARZACHER and SCHNEDL, 1966)

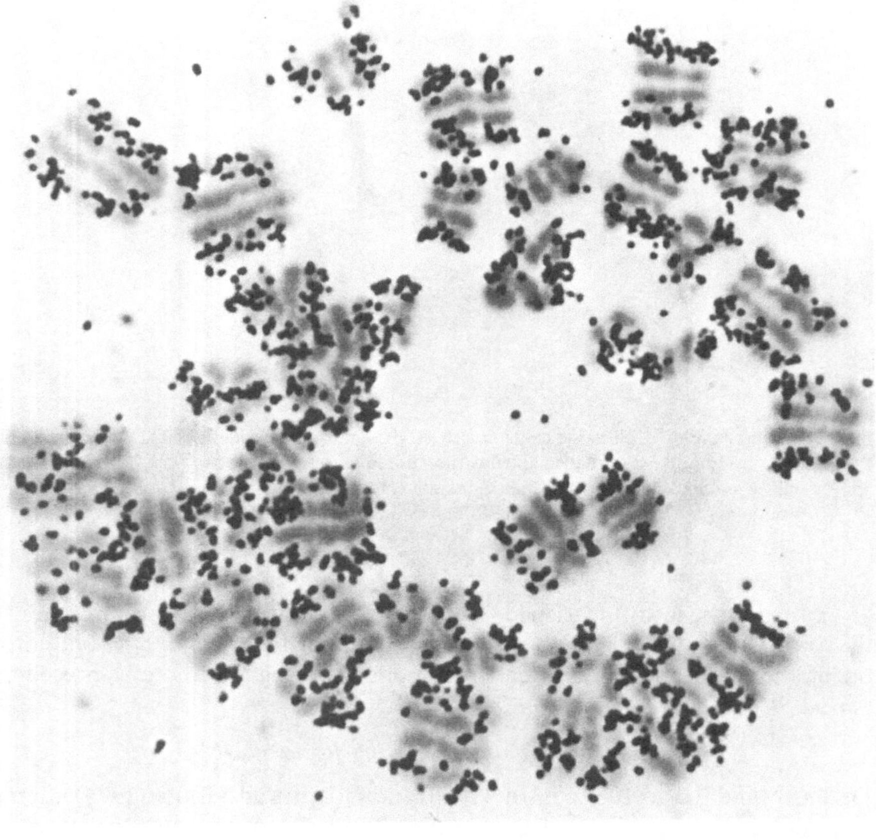

Fig. 69. Metaphase of a human endoreduplicated cell labelled with ^{3}H-thymidin during the first replication cycle. Only the outer chromatids are labelled (compare with Fig. 68)

long and 0.8 μ thick (disregarding the major coils). If a 100 Å thick elementary fibril consisting of a coiled 30 Å fibril is assumed, the packing ratio of DNA could range between 2.1:1 and 7.5:1. Depending on the amount of coiling of the elementary fibre, the packing ratio within the chromatid changes. An assumed length of the elementary fibre of 1 160 μ would give a packing of 28:1, which was first thought to come close to reality. BAHR (1970) suggested later that the packing ratio may be much higher, perhaps as high as 150:1. This would mean that the elementary fibre has to be supercoiled into a fibre of about 450 μ diameter. As has been shown, fibres of this thickness are not found in preparations of directly fixed chromosomes. Recently, GOLOMB and BAHR (1974) published new studies on the total dry mass and the DNA-packing ratio of human interphase nuclei. They come to the conclusion that rather a packing ratio of 28:1 exists as was thought previously. This value would be close to a packing ratio calculated for a chromatin fibre of 200 Å diameter. As was described above, the most probable diameter of the elementary chromatin fibre is, however, around 100 Å. Thus the hitherto reported calculations of the DNA packing ratio from determinations of DNA mass and assumed fibre length are not quite consistent with the electron microscopic results. Obviously more and improved investigations have to be done in this direction.

The calculations of the packing ratio could of course be made in a similar way for a multistranded chromosome model but they show that the packing of DNA into a single strand is possible.

Other Indirect Evidence for Single Strandedness

It should be remembered here that many quite different observations point in the direction of a single stranded chromatid. To mention some: The type of chromatid or chromosome breaks after X-ray irradiation; the theories on mutation and recombination; the calculation by LAIRD (1971) that DNA sequences are present only as single copies (disregarding repetitive sequences) in a haploid set. More detailed information on these topics are available e.g. in the review by COMINGS (1972).

Conclusions

None of the morphological findings can be taken as direct evidence for a multistranded chromatid. They are all compatible with a single stranded model, some of them as for example the observations on prematurely condensed chromosomes or segregation studies are strongly in favour of a single strand. This together with many of the other indirect lines of evidence makes it very probable that a chromatid contains only one DNA double helix, laid in numerous folds, but runing through from one end to the other.

5. Major Coils

Chromosomes in late prophase and metaphase show a coiling (Fig. 70) which can be easily observed in the light microscope (see e.g. DARLINGTON, 1955).

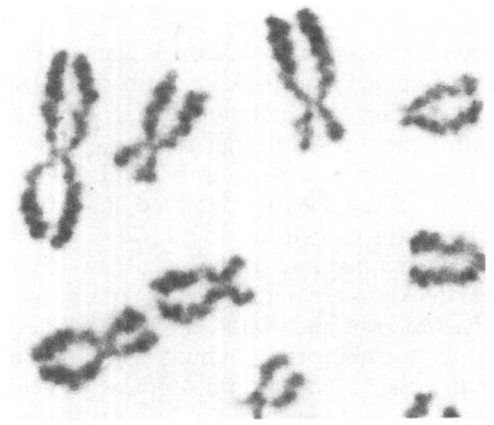

Fig. 70. Human chromosomes in metaphase showing distinctly major coils. Standard chromosome preparation, orcein stain. ×2500

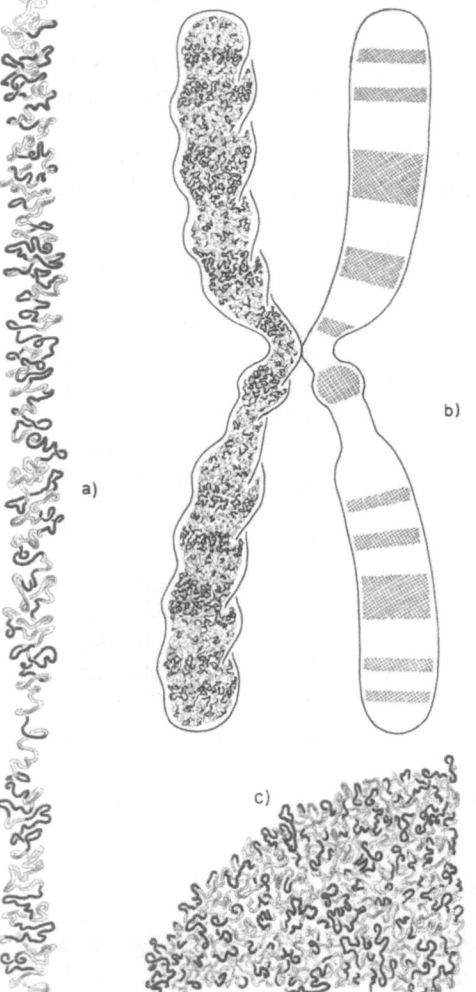

Fig. 71. Schematic drawing of *a*: part of a chromatid in the noncondensed state; *b*: a metaphase chromosome; *c*: part of an interphase nucleus. Represented as they appear after G-band staining. The chromatid (a) is shown at relatively high magnification (compare with Figs. 63–65). The metaphase chromosome (b) is shown at lower magnification. In the left chromatid the major coils and the position of the elementary chromatin fibril are indicated. It is assumed that only one fibril with lightly and heavily stained segments runs through from one end to the other (compare with Figs. 58 and 59). In the right chromatid the G-bands are indicated as seen in the light microscope as a result of the confluence of several smaller darkly stained regions. In the interphase nucleus (c) the non-condensed chromosomes are irregularly packed together (compare with Fig. 73)

In standard cytogenetic preparations, the coiling can be enhanced or made more visible by special pretreatments, e.g. application of hypotonic salt solutions with relatively high potassium ion concentrations (OHNUKI, 1965, 1968). Also dry heat at 100 °C produces a banding after Giemsa staining that corresponds to the position of the major coils. This suggests that the densest parts at the peak of the coils are staining most intensely (COMINGS, 1972).

OHNUCKI (1968) found that the number of major coils is bigger in prophase than in metaphase chromosomes. A detailed study on major coils of human chromosomes was carried out on whole mount electron microscopic preparations by RUZICKA (1973). He found that at a given state of the mitotic cycle, the number of coils is characteristic of a particular chromosome depending on its length. Therefore the number of coils may be used for identification of chromosomes. He found further that in prophase the number of coils is higher for a given chromosome than in metaphase. In some instances the number of coils is twice as high in prophase than in metaphase. The number of coils decreases and at the same time the diameter of the coils increases proportionally during the contraction cycle of the chromosome.

Chromosomes in very early prophase (Fig. 10) or prematurely condensed chromosomes (Figs. 44, 61, and 62) do not show any coiling. The threadlike chromatids appear to be irregularely bent. The main process of shortening and thickening of the chromosomes during pro- and metaphase seems to consist of a simple coiling of the threadlike early prophase chromosomes (see Fig. 71).

6. Giemsa Bands, Interband Zones and Secondary Constrictions

The methods for producing the so called Giemsa bands (G-bands) and the mechanisms involved have already been described in Chapter V. As has also been mentioned, chromosomes treated by means of G-band methods which employ NaOH and concentrated salt solutions at high temperature reveal the banding pattern in electron microscopic preparations as well (RUZICKA and SCHWARZACHER, 1974). In Figs. 58 and 59 electron micrographs of whole mount preparations are shown. In the banding regions the elementary fibrils (here of a diameter of 300–600 Å) are more darkly contrasted but they are also more numerous. Thus, the banding region of the chromatid is thicker than the interband zone. The same can be demonstrated in the thin chromatids of prematurely condensed chromosomes (Fig. 63). Regions known as secondary constrictions appear in whole mount preparations just as zones where the chromatids are thinner. The same elementary fibrils can be found as elsewhere in the chromosomes. Thus, secondary constrictions look very similar to prolonged interband zones. No special structure can be found. The stainability, however, by special staining methods may differ from that of other regions (see Chapter V). One example is the so called "Giemsa-11" method (BOBROW et al., 1972; GAGNÉ et LABERGE, 1972), which stains preferentially the secondary constriction near the centromere of chromosome No. 9. Electron microscopic studies of chromosomes stained with this method demonstrate that the elementary fibrils are very strongly contrasted at this region (Fig. 72).

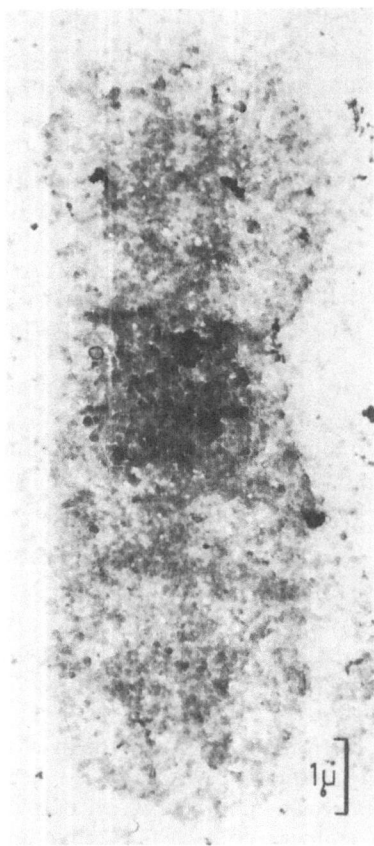

Fig. 72. Electron micrograph of a human chromosome No. 9 after staining with the "Giemsa 11" method of Bobrow *et al.* (1972). The fibrils are strongly contrasted in the centromeric region which appears purple in the light microscope

The correlation of the fine bands to density differences along the chromatids found by Bahr *et al.* (1973) in water surface spread and critical point dried chromosomes has already been discussed. It should be emphazised here that these regularly appearing density differences exist already in the very thin chromatids of prematurely condensed G1 and G2 chromosomes, as well as in early prophase chromosomes (Figs. 63–65). If we take the single elementary fibril model for granted, this would mean that in these regions the fibril is laid in loops of similar appearance to those in the interband regions, but that the band regions are simply thicker. The Giemsa staining effect may cause an additional contrast in the band regions. The similarities between G-bands and chromomeres of giant chromosomes have already been discussed in Chapter V.

7. Bridges Between Chromatids and Chromosomes

Bridges consisting of the same material as the chromosomes themselves between chromatids have been repeatedly reported in the literature (e.g. COMINGS and OKADA, 1972; SHAW *et al.*, 1972; EMMERICH, 1973; JAFFRAY and GENEIX, 1974). These reports are based mainly on observations of whole mount preparations of chromosomes spread on glass or water surfaces. The pictures of such preparations (e.g. Figs. 55 and 56) all show such occasional connections between sister chromatids or between neighbouring chromosomes. At the same time, however, these pictures also show that the chromatids are not sharply outlined. Loops of fibrils stick out from the general border of the chromatids everywhere. Bridges between neighbouring chromatids or chromosomes are therefore easily explained as random connections which have been formed during the course of preparation. All methods of whole mount preparations include a loosening of the fibrillar loops and coils. The fibrils have obviously also a tendency to stick together. To demonstrate this point even more clearly, a picture of three *interphase* nuclei is given in Fig. 74. These nuclei were photographed from the same preparation as Fig. 58. The cells were treated by hypotonic solution, fixed in acetic acid-alcohol, spread on a glass surface, air dried, then stained according to the Giemsa-banding method using NaOH and heat, carbon coated and transferred to an electron microscopic grid. The nuclei have come to lie close together and complete-

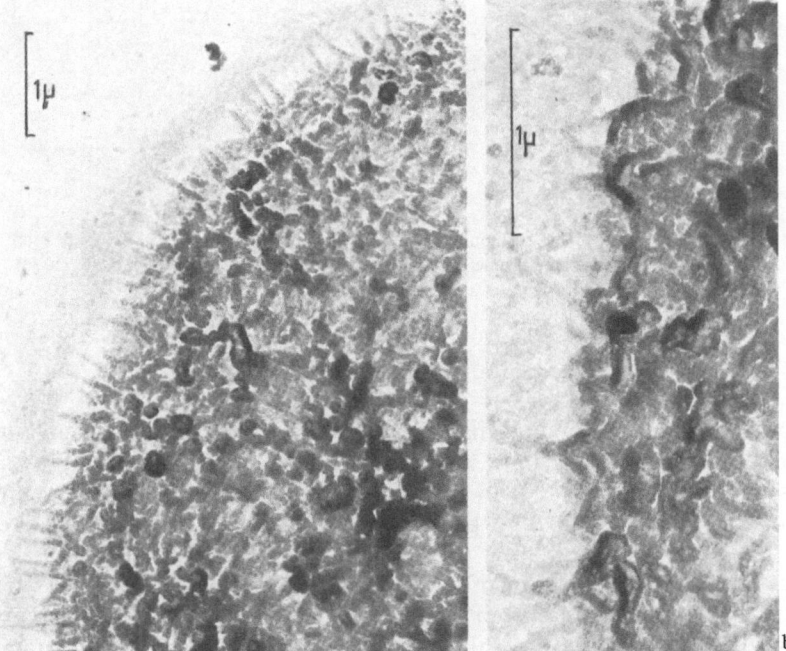

Fig. 73a and b. Electron micrograph from parts of human interphase nuclei. G-banding staining, method as in Fig. 58. Fibrils are irregularely packed. Compare with Fig. 71. (From RUZICKA and SCHWARZACHER, 1974)

ly at random. The same fibrillar bridges are seen between them as between
the sister chromatids and between neighbouring chromosomes in Fig. 58. Nobody
can seriously claim that these fibrils connecting the nuclei represent bridges
which exist in vivo. It seems to be clear that the reports on interchromatid
bridges deal in most cases with artefacts.

A connection of chromosomes in an end-to-end fashion has been described
by WAGENAAR (1969) in certain plants. Only light microscopic studies have been
made so far and a careful re-investigation seems to be necessary in these cases.

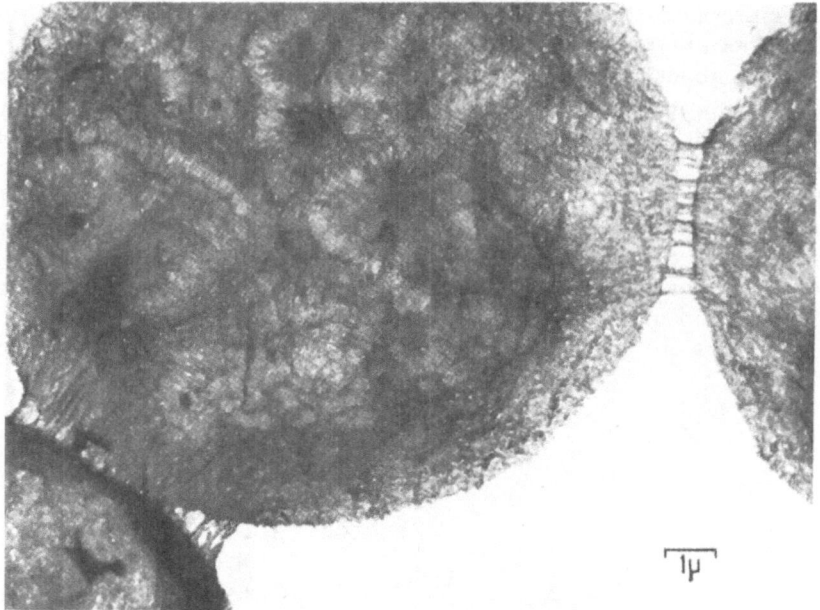

Fig. 74. Electron micrograph of cell nuclei (method as in Figs. 58 and 73). Forming of fibrillar
bridges between nuclei are artefacts. (From RUZICKA and SCHWARZACHER, 1974)

8. Centromeric Region

The centromere is defined as that region of the chromosome where the sister
chromatids stick together until anaphase, and where the spindle fibres attach.
It has long been known that in this region the chromatids are very thin. This
thinning has been called the "primary constriction". In electron microscopic
preparations of many different methods, no particular structure can be seen
at the site of the connection of two sister chromatids (e.g. Figs. 53, 55, and
58). It appears that the chromatids are merely quite thin at this region. As
in secondary constrictions, no difference in the structure of the elementary fibrils
can be observed. This thinning is always seen despite the fact that heterochromatin
(so called centromeric heterochromatin), or secondary constrictions, or nucleolus
organizer sites may be located in close vicinity to the centromere.

COMINGS and OKADA (1970b) proposed the theory that the centromere in non-telomeric chromosomes consists of two attachments in a tandem position. They base their propositions on the observation that in metacentric mouse L-cells the centromeric region is particularly large and shows two thickenings where the chromatids stick together. In telomeric mouse chromosomes only one thickening per chromatid and one attachment region is present. However, also an alternative interpretation of the pictures of COMINGS and OKADA (1970) is possible. The chromosomal regions near the centromere contain heterochromatin and are definitely thicker than other chromosomal regions. What appears as the two attachments are merely the interchromatid bridges formed as an artefact by jutting out loops of fibrils. Since the chromatids are thicker in these regions, they come automatically into close contact. The centromere proper would then be the very thin regions between the thickenings. A similar interpretation can be given for all other chromosomes where thickenings are present on both sides of the centromere. Thus we may conclude that the region where the two chromatids are attached is generally particularly thin, also in terms of electron microscopic dimensions. This means that in this region the elementary fibril is less tightly folded or supercoiled than in the other areas.

The attachment of the spindle fibres is achieved through a special structure, called the *kinetochore*. It consists of two layers of electron dense material in a plate-like form. The diameter of this plate in human chromosomes is a few tenths of a micron. In some special cases it can be as big as 1.4 μ (COMINGS and OKADA, 1971). For a detailed discussion of the structure of the kinetochore see e.g. BRINKLEY and STUBBLEFIELD, 1970; COMINGS, 1972).

The kinetochore cannot be observed in routine cytogenetic preparations leading to an isolation of the chromosomes and to a more or less severe destruction of cytoplasmic structures. EVANS and ROSS (1974) reported, however, dot-like structures in the centromeric regions of total chromosome preparations treated in different ways. These dots may represent kinetochores and their associated proteins.

9. Attachment of Chromosome Fibrils to the Nuclear Membrane

In sections of directly fixed interphase nuclei, the chromatin masses often appear in close association with the nuclear membrane (e.g. HORSTMANN and KNOOP, 1957; ROBBINS and GONATAS, 1965). The hypothesis has been put forward that the chromatin fibrils may be especially attached to the nuclear membrane and moreover, at special sites (COMINGS, 1968; COMINGS and OKADA, 1970c, 1972; COMINGS, 1972). The morphological evidence brought forward for this hypothesis consists solely of pictures from preparations of chromosomes made by the water surface spreading-critical point drying method. In such preparations of interphase nuclei of the honey bee (DU PRAW, 1965b) a particular agglomeration of chromatin fibrils (corresponding to the elementary fibril) at remnants of the nuclear membrane are sometimes seen. This is not surprising, if the general tendency of the fibrils to stick to each other or to any other structure is considered. COMINGS and OKADA (1970c) published pictures of human and Chinese hamster metaphase chromosomes spread and isolated by this method which contain small, rather amorphous looking spots of non-fibrillar material. They claim that these are remnants of the nuclear membrane. One ring-like structure is even interpreted as an annulus of the nuclear membrane. Regarding the very crude mode of preparation and the possibility that all sorts of contamination can occur this interpretation seems rather doubtful.

In electron micrographs of sectioned prophases of HeLa cells (PAWELETZ, 1974) condensing chromosomes can be seen in close vicinity of the still existing nuclear membrane, but no special attachments to any particular site of the nuclear membrane such as nuclear pores are detectable. In sections of metaphases the nuclear membrane is never seen nor any remnants attached to the chromosomes (ROBBINS and GONATAS, 1965; SCHWARZACHER and SCHNEDL, 1967).

10. Summary of the Findings on the Fine Structure of Chromosomes

The basic morphological unit is a fibril of about 20–30 Å in diameter. This fibril represents either the bare DNA molecule or perhaps a DNA-protein complex. The 20–30 Å fibril is folded (a regular coiling cannot be seen with certainty) into a fibril of about 100 Å (80–150 Å) thickness. The 100 Å fibril is called the "elementary fibril". It may reveal periodical thickenings, the so called "v-bodies". They are caused by centrally situated globules of histones around which the DNA molecule winds. Special techniques including hypotonic treatment before fixation, NaOH and heat treatment as for G-banding methods may cause the elementary fibril to thicken, probably by a loosening of the folds of the fine fibril. The elementary fibril therefore measures 200–300 Å in diameter in all preparations where a hypotonic treatment is involved.

The elementary fibrils are laid in irregular folds and loops throughout the chromatid. No distinct longitudinal orientation can be observed. Also, no regular coiling and supercoiling of the elementary fibril is apparent. In the regions of the G-bands, the chromatids are thicker than in the interband regions or in secondary constrictions. At the centromere the chromatids are particularly thin. The centromere probably consists of only one region of sister chromatid attachment. The elementary fibril is laid in folds and loops of the same dimension in the thin regions as in the thickenings of the G-bands. The chromatid is only about 0.5 μ thick in its stretched form in early prophase. Toward metaphase it contracts by a rather regular coiling. The pitches of these major coils remain constant, their number decreases and at the same time their diameter increases accordingly.

The question as to whether only a single elementary fibril running through from one end to the other forms a chromatid is not yet quite decided. Although almost all findings favour this assumption, the possibility that several elementary fibrils build up a chromatid cannot be completely ruled out.

Fibrils crossing from one chromatid to the other or between neighbouring chromosomes must be considered as artefacts. No morphological evidence exists for the attachment of chromosomes to the nuclear membrane.

In Fig. 71 a model of a metaphase chromosome is given based on the assumption that one elementary fibril is present per chromatid.

VII. Chromosome Structure in Interphase Nuclei

1. Introduction

In the foregoing chapter, the main emphasis was placed on the structure of chromosomes in mitosis. Chromosomes contract with the onset of mitosis forming microscopically visible single thread- or rodlike structures. After mitosis is completed, the daughter nuclei are restored involving a disappearance of the chromosomes as individual microscopic bodies. During the interphase the mass of chromosomes forms the so-called *"chromatin"*. Since the chromosomes reappear in each mitosis in exactly the same shape and number (even chromosome rearrangements, breaks and other abnormalities are carried through interphase from mitosis to mitosis), it seems to be quite clear that they keep their individuality during interphase. It seems to be also apparent that they are extended into rather thin and long threads in interphase. The best direct evidence for the persistence of the chromosomes comes from observations on polytene chromosomes (as e.g. in the giant cells of salivary glands of Drosophila) during interphase (HEITZ and BAUER, 1933; see e.g. BEERMANN, 1972; HENNIG, 1974).

Besides chromosomes and nucleoli, no constant structural elements have been found in interphase nuclei. The only exceptions are the so called sphaeroids or nuclear bodies (HORSTMANN, 1965; HORSTMANN et al., 1966) which are occasionally found in certain tissues, and whose nature is unclear as yet. The electron microscope also reveals a varying number of small particles which do not necessarily belong to chromosomes and are probably RNA-particles (BERNHARD, 1967).

During interphase, the chromosomes go through the three main periods of the cell cycle. Most morphological investigations of chromatin structure have ignored these periods. Therefore the results of such investigations and the following review of them are valid only for interphase as a whole. Possible changes of nuclear structures during the cell cycle will be discussed in an extra paragraph.

2. Structure of Interphase Nuclei of Living Cells

Since the first observations of nuclear structure in living cells (LEWIS, 1923; CHAMBERS, 1924) it is known that cell nuclei appear rather clear and homogeneous. Only the nucleoli and the nuclear membrane are distinctly visible in the light microscope. Later investigations with the aid of the phase contrast microscope revealed that nuclei in the living state appear to have a rather opaque and homogeneous content with occasional cloudy condensations (e.g. STOCKINGER, 1953; ALTMANN and GRUNDMANN, 1955). In some tissues of certain

mammalian species small bodies of condensed chromatin, the so-called chromocenters, are visible. They correspond to heterochromatic regions and will be
discussed in detail later. Exact descriptions of chromocenters in supravital liver
cell nuclei of the mouse are given by PISCHINGER (1937).

In human cell nuclei no larger chromocenters can be seen in living or unfixed
cell nuclei. Only the X-chromatin body (Barr body, sex chromatin body) of
female cells can be observed in a certain percentage of nuclei of living fibroblast
cultures (DE MARS, 1962; SCHWARZACHER, 1963, see Fig. 75).

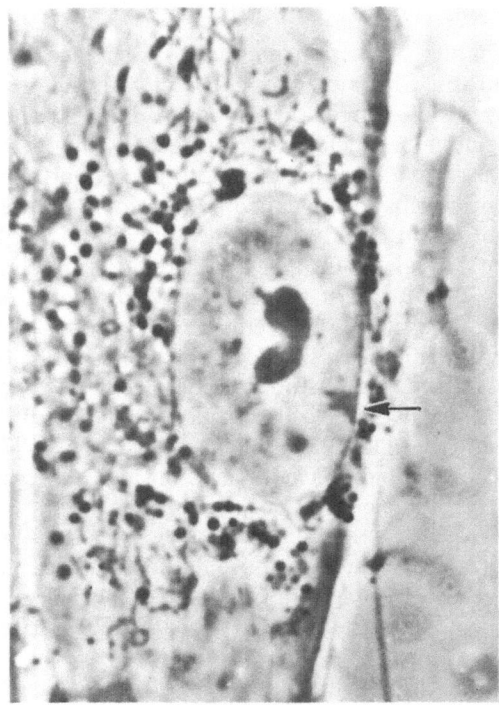

Fig. 75. Living cell from a human fibroblast culture. The arrow points to the X-chromatin. Phase
contrast, × 1 800

3. Structure of Interphase Nuclei after Fixation

Most fixatives used in histology and cytology cause an agglomeration of chromatin into small bodies and threads, leading to a characteristic nuclear structure.

It has long been known that cells of different species and tissues can be
recognized in fixed preparations to a certain degree by their nuclear structure
(HEIDENHAIN, 1907). This means that in a given cell, the homogeneous chromatin
is altered by the fixation in a constant way. There is no doubt that with standardized methods of fixation different cell types can be differentiated solely by
their nuclear structure (GRUNDMANN and STEIN, 1961; MÜLLER, 1966). These
constant differences are caused by a different agglomeration and coacervation

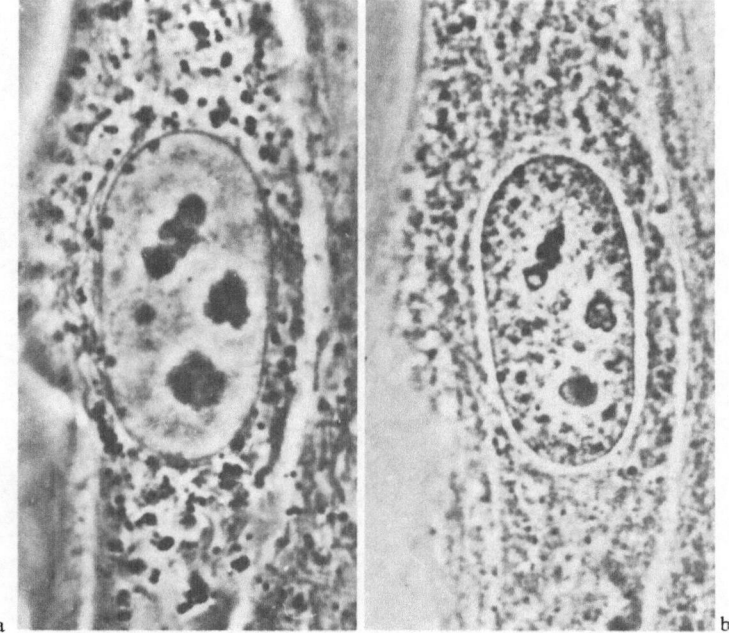

Fig. 76a and b. The same cell from a human fibroblast culture (a) in the living state, (b) after fixation in 95% ethanol for 5 min

of the chromatin substance which depends obviously on its physico-chemical state. On the other hand the type of fixation is of primary importance, since different fixatives produce quite different appearences of the nuclei.

The alterations of the nuclear structure by different fixatives were studied in detail by ZEIGER (1935, 1938) and by PISCHINGER (1937, 1950). Acid fixatives lead to dehydratisation and coacervation of the nucleoproteins causing a coarse granulated appearance of the chromatin. Ethyl and methyl alcohol act in a similar manner (Fig. 76). Aldehydes as fixatives lead to a much finer granulated appearance, glutaraldehyde in particular preserving the nuclear structure almost as it appears in the living state. The same preservation of the *in vivo* appearance can be achieved by fixation with heavy metal ions (Fig. 77). Both types, glutaraldehyde and osmic acid, are at present the fixatives of choice in electron microscopy. Osmic acid produces particularly homogeneous looking cell nuclei. This may be due to the hypotonic effect of this type of fixation. If surviving cells are treated with hypotonic salt solutions, the nuclei become very clear and even chromocenters, if present, disappear (Fig. 78). This effect is reversible. Hypertonic solutions cause a granulated appearance, an effect which is likewise reversible.

Several ways have been tried to describe and classify the structural differences after fixation. GRUNDMANN and STEIN (1961) attempted this on unstained cells after fixation in alcohol-acetic acid, by classifying subjectively the number and size of coarser chromatin particles. SANDRITTER *et al.* (1967) developed a photo-

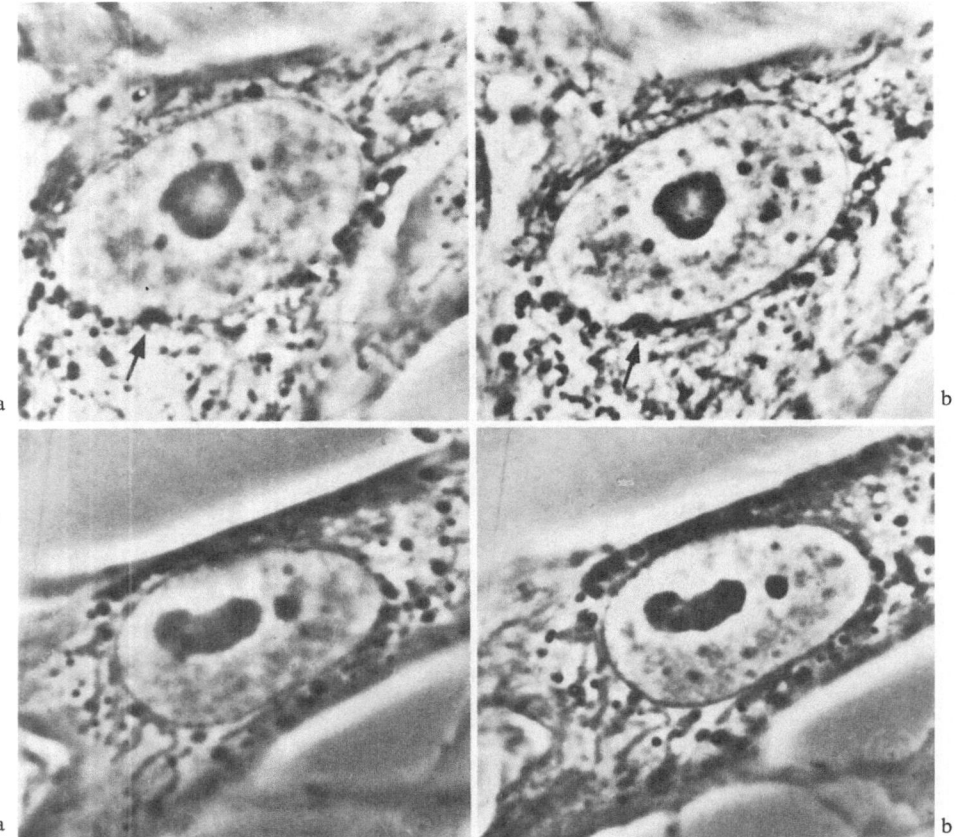

Fig. 77a and b. Cells from a human fibroblast culture, (a) in the living state and (b) after fixation in buffered osmic acid (pH = 7.4, after PALADE) for 30 min. (From SCHWARZACHER, 1964). Courtesy of S. KARGER, publishers

metric method on standardized Feulgen stained nuclei. The chromatin particles were classified according to their degree of extinction. At least three classes of nuclei could be easily discerned in this way.

4. Chromocenters

In addition to the more or less fine granules which appear in the chromatin after fixation, the previously mentioned constantly occurring bodies of condensed chromatin, called chromocenters, exist. TELLYESNICZKY (1905), who described them for the first time in animal nuclei, called them "karyosomes". They are identical with the "chromocenters" described by ROSENBERG (1904, 1909) in plant cell nuclei. ROSENBERG was already aware that they were of chromosomal origin. On the basis of these earlier studies HEITZ (1923, 1929) showed that the chromocenters correspond to certain parts of the chromosomes which do not decondense during interphase. From these observations he developed his concept of heterochromatin. We may rather formulate: Heterochromatic chromosomes or chromosome regions remain more strongly condensed during interphase than the rest of the euchromatic chromosomes. This property of heterochromatin is called heteropyknosis (see also Chapter VIII). Since the amount of heterochromatin per genome differs widely from species to species, the amount and the size of chromocenters in different interphase nuclei vary considerably too.

In human cells the most important chromocenter is formed by the heterochromatic X-chromosome in female cells (Barr-body, X-chromatin). In general the X-chromatin reacts like the other chromatin against different fixatives. It becomes coarser with acid fixatives, but is hardly altered with aldehydes. It also becomes reversibly less dense in hypotonic solutions and reversibly denser in hypertonic solutions (SCHWARZACHER, 1964, see Fig. 78). Thus in general, the rather crude influence of histological fixation or hypo- and hypertonic treat-

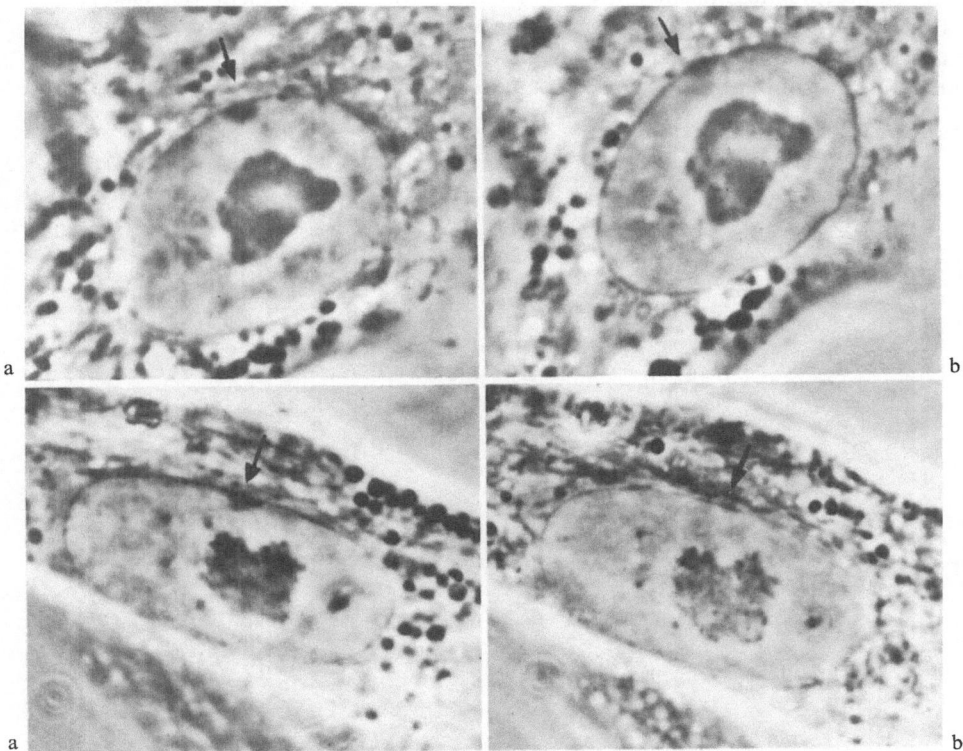

Fig. 78a and b. Cells from female human fibroblast culture in the living state (a) with distinct X-chromatin (arrows) and (b) after treatment with hypotonic solution (Hanks: dest. water 1:3) for 5 min. The X-chromatin has disappeared in (b). (From SCHWARZACHER, 1964. Courtesy of S. KARGER, publishers)

ment on chromocenters is the same as that on the rest of the chromatin. There are, however, also observations of a differential behavior between heterochromatin and euchromatin under certain circumstances. This shall be described in context with a short review on heterochromatin and euchromatin in Chapter VIII.

5. Electron Microscopy of Chromosomes in Interphase

In general the same types of fibrils as in mitotic chromosomes can be found in interphase nuclei. A comparison of the chromatin structure with metaphase chromosomes demonstrates that the same structural elements are present in both but that the degree of condensation is different.

As has already been mentioned, thin sections of directly fixed glutaraldehyde and osmic acid material give the most reliable preparations. In Figs. 79 and 80 sections of interphase nuclei (of human fibroblast cultures) are shown. At higher magnification it can be seen that the impression of a fine granulation is in reality caused by sections through fibrils which are bent and folded in very short loops. The diameter of the fibrils is about 100 Å; they thus have the same thickness as in mitotic chromosomes, and represent the elementary fibril. As in mitotic chromosomes fine fibrils of about 20–30 Å thickness can also be seen. So far all investigations on interphase nuclei in different species have given similar results (e.g. HORSTMANN and KNOOP, 1957; BARNICOT and

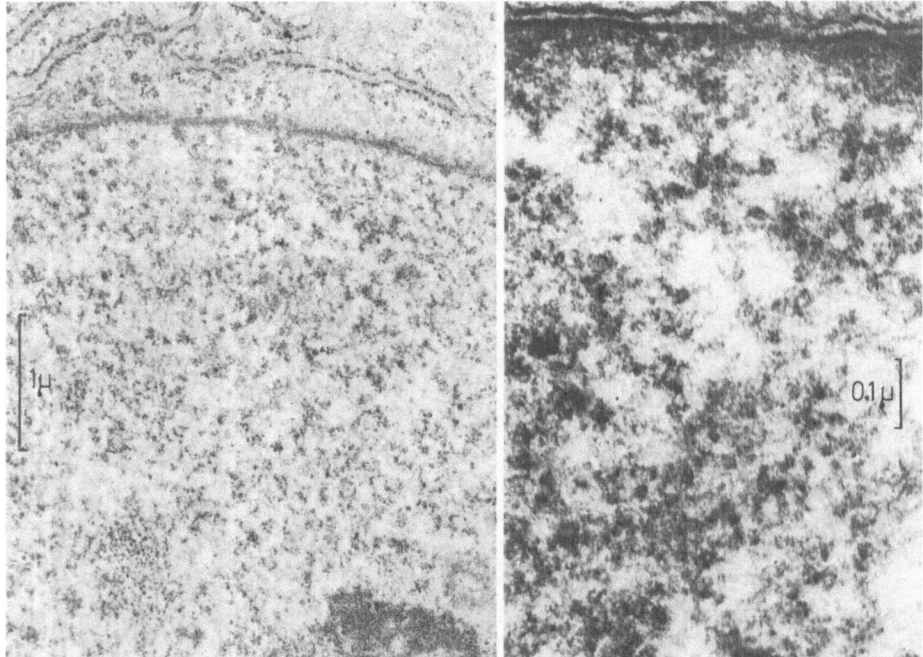

Fig. 79

Fig. 80

Figs. 79 and 80. Electron micrographs of parts of interphase nuclei from human fibroblast cultures

HUXLEY, 1961; HAY and REVEL, 1963a; SCHWARZACHER and SCHNEDL, 1967). The fibrils are usually not as tightly packed in interphase nuclei as in metaphase chromosomes. However, in regions of chromocenters, a rather tight packing is observed, the fibrils themselves not differing from those of euchromatic chromatin regions (WOLSTENHOLME, 1967).

Particularly clear pictures of interphase chromatin fibrils were published by HEUMANN (1974). Most remarkably HEUMANN was able to demonstrate RNA around the chromatin fibrils along certain stretches, interspersed by RNA-free regions.

With different techniques the appearance of the fibrils alters in just the same way as in mitotic chromosomes. Whole mount preparations without any special treatment are usually too thick to reveal any details. But after spreading of the cells on a liquid surface and critical point drying the same sorts of fibrils are found as in metaphase chromosomes (GOLOMB and BAHR, 1974; TANAKA and IINO, 1974).

Whole mount preparations treated as for G-banding techniques (RUZICKA and SCHWARZACHER, 1974) also show fibrils in interphase nuclei which correspond exactly to those in metaphase chromosomes (Fig. 73). The diameter of the elementary fibril in this case is at least 300 Å. In addition, fine fibrils building up the elementary fibril can be detected at high magnification (see Chapter VI).

In G-banding treated preparations two points deserve particular attention: 1. The fibrils are laid in similar loops and foldings as in metaphase chromosomes, although they are obviously not so tightly packed. 2. There are small dark regions in which the fibrils are more densely packed as well as being thicker and more heavily contrasted. These latter regions presumably correspond to the G-band regions of mitotic chromosomes. Thus, the banding of the chromosomes persists through interphase.

In Chapter VI a model of a slightly contracted chromatid is proposed (Fig. 71a). Assuming that in interphase the chromosomes are slightly less condensed and irregularly folded and twisted together within the nucleus, a picture is obtained, which comes quite near to the real electron microscopic picture (Fig. 71c).

6. Premature by Condensed Chromosomes

Further information on the structure of chromosomes in interphase has become available by the study of prematurely condensed chromosomes. As has already been explained in Chapter VI, the condensation of interphase chromosomes can be achieved by fusing cells in metaphase with cells in interphase (Fig. 61). It seems that the cell which is farther advanced in the cell cycle forces the less advanced cell to synchronize with it. In the same way, a cell in mitosis induces condensation of the chromosomes in a cell which may be either in G1, S or G2 (RAO and JOHNSON, 1970; JOHNSON and RAO, 1970; MATSUI et al., 1972). It is so far not clear, by what factors the premature chromosome condensation is induced (see SPERLING and RAO, 1974).

Fig. 81. Part of a fusion figure between a HeLa cell in metaphase and a cell in G1. Premature condensed chromosomes thin and single stranded. ×2400

Fig. 82. Part of a fusion figure between a HeLa cell in metaphase and a cell in G2. Premature condensed chromosomes consist of two chromatids. ×2400

In the light microscope it is possible to differentiate between prematurely condensed chromosomes of different stages of the cell cycle in hypotonically pretreated and air dried preparations. In Fig. 81 part of a fusion figure is shown. The prematurely condensed chromosomes are very thin threads. Since they consist of only one chromatid, they are obviously G1-chromosomes. In Fig. 82 the prematurely condensed chromosomes consist of two fine threads. They were obviously in G2 when induced to undergo premature condensation. In both G1 and G2-prematurely condensed chromosomes, all the structural details as seen in prophase and metaphase chromosomes can be detected. The centromeric regions are usually well seen and also secondary constrictions and satellites are observable. This merely gives proof of the selfevident fact, that the chromosomes persist through interphase with all their structural features.

After G-band staining, a very fine banding pattern can be observed (UNAKUL et al., 1973; SCHWARZACHER et al., 1974). As has been already pointed out, there are many more bands to be seen than even in prophase chromosomes, but the bands are of certain minimum size and are divided by negatively stained regions (Figs. 44 and 45). From the size and number of the G-bands alone, a possible correspondence to the chromomeres of giant chromosomes may be suggested (see Chapter V). The electron microscopic pictures (Figs. 62 and 64) again show the G-band regions as more densely packed and with more strongly stained fibrils.

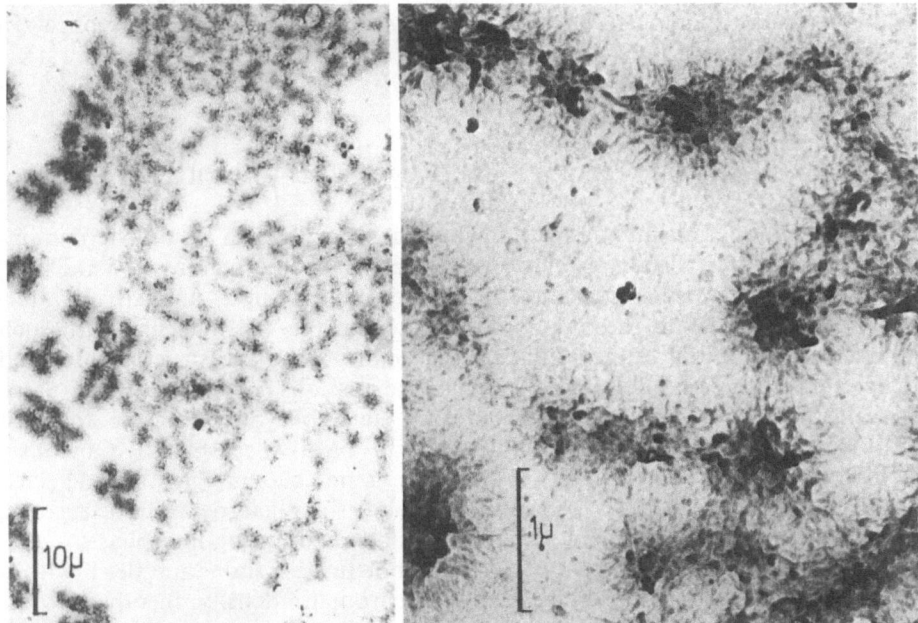

Fig. 83

Fig. 84

Figs. 83 and 84. Electron micrographs of premature condensed human chromosomes in S-phase. The fibrils between the condensed chromatid regions are presumably formed by sticking out loops in most cases

Another point should be taken into consideration when studying the electron micrographic pictures of very thin G1-chromosomes (Fig. 63): The degree of condensation is rather low, when compared with interphase nuclei, prepared in the same way (Fig. 73). This means that the premature condensation consists not so much of a dramatic condensation of the fibrillar loops but rather of a separation of the chromosomes from each other.

Interesting is the situation if chromosomes in the S-phase are brought to premature condensation (Fig. 83). Such chromosomes consist of many small bodies. This has been called "fragmentation", but it resembles the condensed chromosome parts which are not replicating, whereas the replicating DNA is situated between the condensed parts (RAO and JOHNSON, 1974). In the state of replication the DNA is presumably maximally uncoiled and of course not visible in the light microscope. But even electron microscopy of prematurely condensed chromosomes in S-phase has so far not succeeded in demonstrating the replicating DNA molecule (SCHWARZACHER et al., 1974). In Fig. 83 an electron microscopic picture is given of prematurely condensed chromosomes in S-phase at low magnification. The preparation was done using a standard chromosome technique, the G-banding staining, and then whole mount transfer onto an electron-microscopic grid. At higher magnification (Fig. 84) there are elementary fibrils clearly visible in the condensed parts of the chromosomes. There are also bridges of fibrils between the condensed parts, but more probably these are artefacts formed by fibrillar loops sticking together. No thin fibrils between the condensed parts can be found. Thus with the present methods available the morphological state of the DNA-protein fibril during replication is still obscure.

7. Special Chromosome Regions in Interphase

In Chapter IV special staining methods were described which are able to demonstrate differentially certain chromosomes or chromosome regions (C-bands, Q-bands, G-bands, Giemsa-11-bands etc.). The same methods applied on interphase nuclei may stain these band regions within the cell nucleus. With such methods we should have the means to learn more about the structure assumed by the chromosomes in interphase. None of the methods have, however, revealed chromosomes in a completely decondensed state, but rather in the form of more or less condensed bodies or granules. This has two reasons: 1. Obviously a small region in a decondensed chromosome is not big enough to be detected in the microscope. 2. Most of the special methods stain heterochromatic regions, which in general have the tendency to stay condensed during interphase.

Besides the special staining methods, common nuclear stains and the Feulgen-reaction may demonstrate regions of higher chromatin density, in other words, the chromocenters. If it is known, which chromosome or chromosomal region is responsible for the formation of a chromocenter, its behaviour in interphase can be studied. In human cells this is especially the case with the heterochromatic X-chromosome in females, which forms the X-chromatin.

Quinacrine Staining and the Y-Chromatin

As has been described in Chapter IV, the Y-chromosome usually contains the largest block of quinacrine positive material of the whole human genome (Fig. 26). About the distal half of the long arms of the Y give a very brilliant quinacrine fluorescence. The largest fluorescing chromatin corpuscle in interphase corresponds therefore to this part of the Y-chromosome (ZECH, 1969; PEARSON *et al.*, 1970). It has been termed the "Y-chromatin" (also "F-body" or "Y-body").

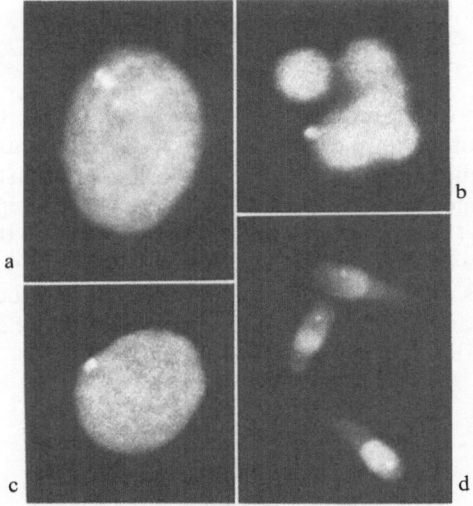

Fig. 85a–d. Quinacrine mustard staining of different cell nuclei of normal men. The Y-chromatin well visible in all instances. (a) Buccal smears. (b) Granulocyte. (c) Large lymphocyte. (d) Sperms. ×2400. (From SCHWARZACHER 1974)

The Y-chromatin is of very small size, usually not bigger than 0.5 μm but thanks to its brilliant fluorescence it is readily seen in almost every Y-chromosome bearing cell nucleus (Fig. 85). In normal cases this means in male cells. Only if the Y-chromosome is particularly small, which is a normal but rare variant (see Chapter IV), the Y-chromatin body might be so small that it could be confused with other fluorescing chromosome regions. In well prepared blood smears of normal males, the percentage of lymphocytes with a distinctly visible Y-chromatin is over 90. In sperms, values of 40–48% have been published (PEARSON and BOBROW, 1970; BARLOW and VOSA, 1970; SUMNER *et al.*, 1971). The quinacrine fluorescence staining of interphase nuclei has therefore some practical value. Besides a simple investigation of the chromosomal sex, it is possible to detect abnormalities of the Y-chromosome (e.g. the diagnosis of individuals with two Y-chromosomes-47 XYY) just by means of studying interphase nuclei. Suitable material for such studies can be easily obtained from patients, for instance oral mucosa smears, hair root cells, or blood cells (see e.g. SCHWARZACHER, 1974 for problems on the technique).

The Y-chromatin is always seen as a corpuscle, not as a thread. This means that the distal part of the long arm of the Y-chromosome is not extended, but condensed in interphase. The Y-chromatin can also be seen after Feulgen staining as a small dense granule, but an exact identification is only possible by a comparison with the quinacrine staining.

Other prominent fluorescing chromosome parts besides the Y-chromosome can sometimes also be seen in quinacrine stained interphase nuclei. The secondary constrictions in the short arms of the nearly acrocentric chromosomes or in chromosome No. 3 can achieve such a size that they may even be confused with a small Y-chromatin body.

C-Band Staining

Most human chromosomes contain C-band material near the centromere and some secondary constrictions are particularly strongly stained by the C-band method, as in chromosomes Nos. 1, 9, and 16, and the distal part of the long arm of the Y-chromosome (see Chapter IV). All these regions are seen also in interphase nuclei (Fig. 86) as small dots. Again none of these regions appear as thin threads as would be expected if they were decondensed. The dot-like appearance indicates the relatively strong condensation of these heterochromatic regions. With the aid of variations of the C-band technique some of the C-band regions can be visualized specifically (KIM, 1974).

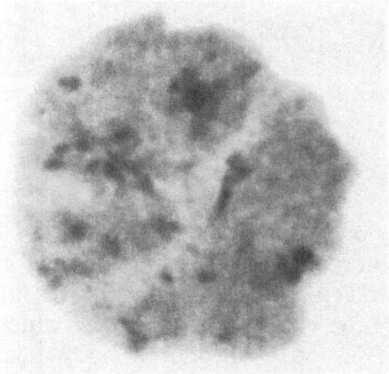

Fig. 86. Cell nucleus from a human lymphocyte culture. Staining of constitutive heterochromatin according to the method of ARRIGHI and HSU (1971). × 2500

G-Band Staining

Interphase nuclei treated according to the G-band staining techniques show numerous darkly stained very fine granules (Fig. 87). From what is known of the G-bands in prematurely condensed chromosomes (Fig. 44), they must be rather small in interphase. The bands are certainly too fine and numerous

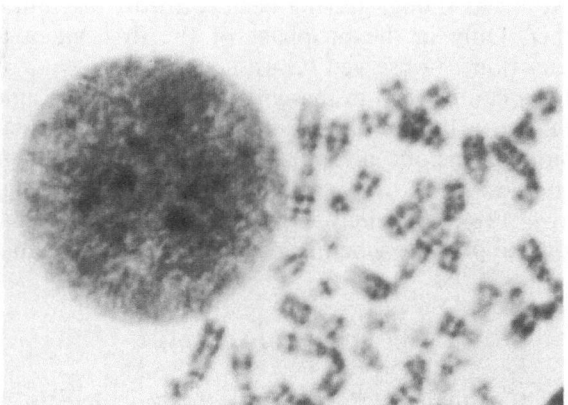

Fig. 87. Cell nucleus and part of a metaphase from a human lymphocyte culture. G-banding staining according to the method of SCHNEDL (1971). ×2500

to allow any specific identification in interphase nuclei in the light microscope. In the electron microscope the band regions are clearly visible. They consist of more densely packed and more strongly contrasted fibrils (Fig. 73). It is, however, hardly possible to identify any specific band in an interphase nucleus.

Giemsa-11 Staining

With this method (see Chapter IV) the secondary constriction near the centromere of chromosome No. 9 is especially strongly stained (Fig. 20). Sometimes other heterochromatic regions are also stained, but usually not as deeply as chromosome No. 9. In Fig. 88 interphase nuclei are shown, stained by this method. The two dark bodies (appearing red against the blue background of the rest of the nucleus) correspond to the No. 9 chromosomes. The form of the bodies is again bulky, indicating the relatively strong condensation of this chromosomal region during interphase. Sometimes the bodies reveal a somewhat

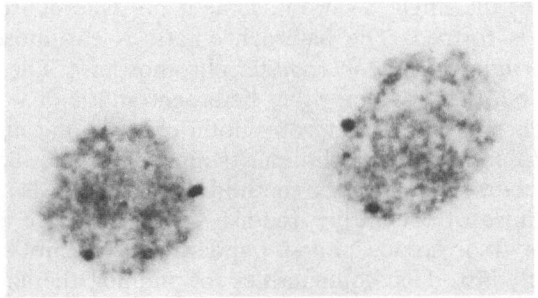

Fig. 88. Two cell nuclei from a human lymphocyte culture. Staining with "Giemsa-11" after BOBROW et al. (1972). The heterochromatic regions of the two chromosomes No. 9 appear purple in the microscope (black and white photographe using a green filter). ×2000

rod-like structure, or as though having been formed by a thinner thread which is folded together. Only in the prophase of the first meiotic division can a certain decondensation be observed (GAGNE et al., 1974).

The No. 9-body can be seen in many different cell types, and also in sperms (BOBROW and PEARSON, 1972). It has been used to try to calculate the frequency of non-disjunctions in meiotic divisions by counting the percentage of sperms with two No. 9 bodies. PAWLOWITZKY and PEARSON (1972) published surprisingly high numbers (over 1%), which may, however, not remain unchallenged. Technical errors seem to be possible when using this rather tricky Giemsa-11 method.

8. The X-chromatin

The discovery of the small heterochromatic body in female mammalian cells by BARR and BERTRAM (1949) and its interpretation as being derived from one of the two X-chromosomes by OHNO and coworkers (see OHNO et al., 1961; OHNO and MAKINO, 1961; OHNO, 1967) were doubtless the starting points for the recent development of human cytogenetics. Many reviews and books exist on the X-chromatin, also called "Barr-body" or "Sex-chromatin" (e.g. BARR, 1966; MOORE, 1966; MITTWOCH, 1967). Therefore only some of the important facts shall be briefly reviewed here, as well as some more recently published results. The study of the X-chromatin has also its practical value in clinical medicine. For information on this aspect of the problem the reader is referred to handbooks on Human Genetics (e.g. HAMERTON, 1971) and to methodological books (e.g. YUNIS, 1964; SCHWARZACHER and WOLF, 1974).

Definition

The X-chromatin is a small dense chromatin particle, seen normally only in cell nuclei of female individuals (Fig. 89). In polymorph nucleated leucocytes the X-chromatin may form an extra small appendage, called *Drumstick* (DAVIDSON and SMITH, 1954, see Fig. 92). It is an X-chromosome which is strongly condensed during interphase and genetically inactive (LYON, 1962). In diploid cells with two X-chromosomes only one is heterochromatic. In normal diploid male cells (46, XY) the single X-chromosome is not heterochromatic and hence no X-chromatin is formed. The heterochromatic X-chromosome is a typical example of a *facultative heterochromatic* chromosome: The X-chromosomes are homologues, but they may be either heterochromatic or not.

Since the X-chromatin is a heteropycnotic chromosome it is demonstrable in the light microscope by all techniques staining chromatin. A very reliable technique is of course the Feulgen-method for DNA. Also acridine orange can be used because of the rather reddish fluorescence of condensed DNA against the green fluorescence of less condensed isopycnotic DNA (RIGLER, 1966). The recently introduced quinacrines for staining the chromosome bands can, besides staining the Y-chromatin brilliantly, also stain the X-chromatin because of its higher density (WYANDT and HECHT, 1971; IORIO and WYANDT, 1973). With careful staining it is quite possible to differentiate between Y-

and X-chromatin by quinacrine fluorescence. For detailed studies, however, one of the specially developed nuclear stains (see e.g. SCHWARZACHER, 1974) are to be preferred. For the demonstration of the drumstick in polymorph nucleated leucocytes a standard stain for blood films (e.g. Giemsa) is adequate (see TOLKSDORF, 1974).

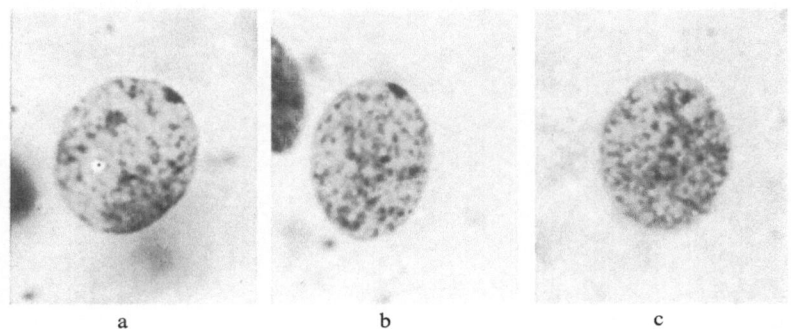

Fig. 89a–c. Cell nuclei from buccal smears, stained with carbol fuchsin. (a) and (b) Normal adult female. (c) Normal adult male. ×2000. (From SCHWARZACHER, 1974)

Form

In most cell nuclei the form of the X-chromatin is triangular, or it seems to be built up of two small rodlets (Figs. 89–91). The latter configuration comes presumably from one small rodlet or thread which is bent approximately in the middle (KLINGER, 1957). A special form is the "*drumstick*" in polymorph nuclear leucocytes (Fig. 92, see e.g. MITTWOCH, 1967; TOLKSDORF, 1974).

Position

The position of the X-chromatin within the nucleus is in many tissues characteristically at the periphery, very often attached to the nuclear membrane. In nerve cells it is more commonly situated in the center of the nucleus. In polymorph nucleated leucocytes it may form the already mentioned drumstick (Fig. 92).

Size

The size, depending on its form, is about 1–2 μm. Size depends also on the cell cycle: in cell nuclei in the G2-period it is about twice as big as in G1-nuclei. In tetraploid nuclei either one double sized X-chromatin or two normal sized X-chromatin bodies are present according to the state of the chromosomes (endoreduplicated cells, or with a random chromosome distribution, see Fig. 93). In higher polyploid nuclei, X-chromatin bodies are formed according to the rule that for each diploid female chromosome set (46, XX) one X-chromosome is heterochromatic and produces one X-chromatin mass (KLINGER and SCHWARZACHER, 1960).

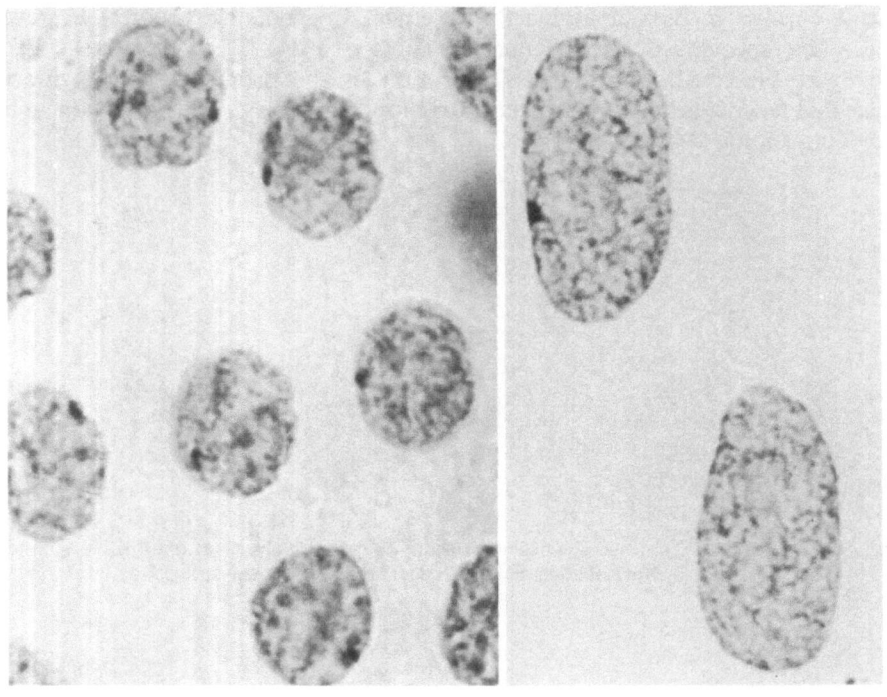

Fig. 90   Fig. 91

Fig. 90. Cell nuclei of the amnion epithel of a newborn girl. An X-chromatin visible in each nucleus. Feulgen stain. × 2000

Fig. 91. Two cell nuclei of a human female fibroblast culture. One X-chromatin positive and one negative nucleus. Feulgen stain. × 2000. (From SCHWARZACHER and PERA, 1970)

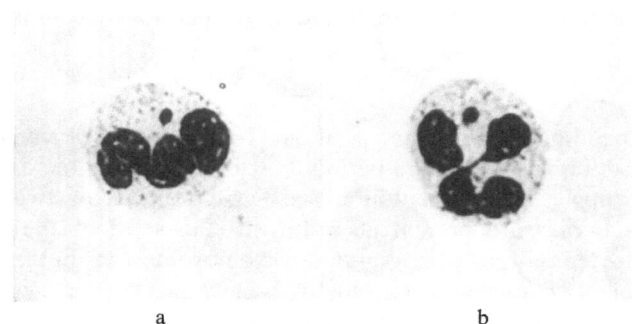

a   b

Fig. 92a and b. Female granulocytes from a blood smear with a typical drumstick on a thin stalk. May-Grünwald-Giemsa stain. × 1600. (From TOLKSDORF, 1974)

From form and size, we can conclude that at least a large portion of the X-chromosome if not the entire chromosome is heterochromatic. The figure of a double rodlet suggests a bending at the centromere region. Compared to the heterochromatic part of the Y-chromosome, the X-chromatin is at least 5 times bigger. A comparison with the Giemsa-11 body of chromosome No. 9 is not possible, since this method leads to a strong swelling of the chromosomes.

Frequency

In a given preparation of a normal female tissue, not all cell nuclei will contain an X-chromatin body. The frequency of the occurrence varies from tissue to tissue (see Figs. 90 and 91). These differences may have several reasons:

Technical factors may play an important role. In smear preparations of oral mucosa cells from female persons, only 25–50% of the nuclei are X-chromatin positive. This low percentage is certainly partly due to technical difficulties in producing smear preparations. Also, the type of fixative is important: in fibroblast cultures from females (growing in the logarithmic phase) about 40–50% of the cell nuclei show an X-chromatin when observed in living cells in the phase contrast microscope (SCHWARZACHER, 1963). After fixation in 95% alcohol, this frequency may go up to 60–70%, because the fixative causes agglomeration of the chromatin. In this way, X-chromatin bodies which are for some reasons less condensed and invisible in unfixed cells may become denser and sharply outlined with fixation. Statements on the frequency of X-chromatin positive cells in the literature have therefore to be interpreted with some caution. For a further and detailed discussion on technical factors see MOORE (1966) and SCHWARZACHER (1974).

The state of metabolic *activity* of the cell seems also to exert an influence. KLINGER et al. (1968) found that in cultured cells the percentage of X-chromatin positive cells increases with the cell density. The cell density is on the other hand related to activity. Another example in this direction was given by SCHNEIDER et al. (1973) on the chromocenters of *Microtus agrestis*. Although the chromocenters of *Microtus agrestis* are for the most part constitutive heterochromatin, this study may be relevant to the facultative X-heterochromatin. It was found that in stimulated thyreoid cells, the chromocenters are not seen, or are only faintly visible, whereas in inactive cells they are quite distinct. Metabolically active cell nuclei very often show an increase in size ("functional swelling"). This is probably connected with an uptake of water and therefore with alterations of the physicochemical conditions of the nuclear chromatin, which in turn may lead to a changed degree of condensation of the heterochromatin.

The *cell cycle* does not seem to have a striking influence on the microscopical visibility of the X-chromatin. Thus, if in a given cell the heterochromatic X-chromosome is distinctly heteropycnotic, it stays so during the whole cell cycle. KLINGER et al. (1967) and MITTWOCH (1967) found no evidence that the X-chromatin decondenses with DNA-replication in the S-period. Heterochromatin in general may, however, show certain morphologic alterations shortly before mitosis. In plant cells a dissolution of chromocenters at the end of the G2-period

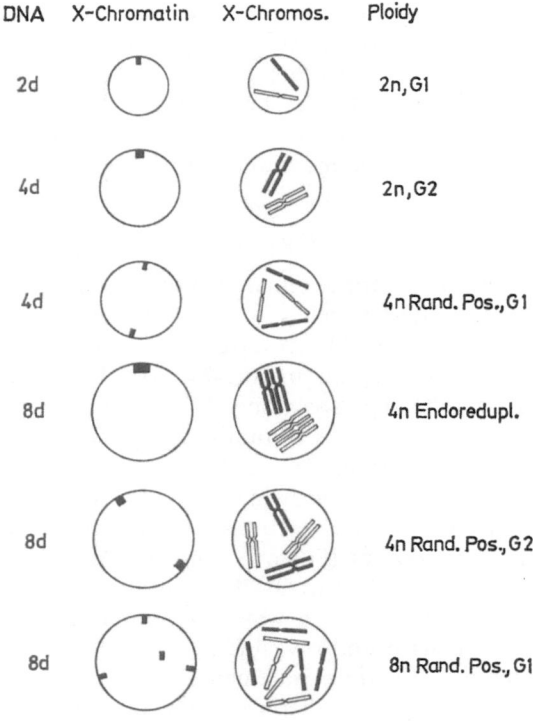

Fig. 93. Diagrammatic representation of the forms in which X-chromatin bodies appear in cell nuclei of various classes of DNA (*2d, 4d, 8d*) and their chromosome status. *Rand. Pos.* Random arrangement of chromosomes. *Endoredupl.* State according to endoreduplication. Heterocyclic chromosomes black. (From SCHWARZACHER, 1974)

is a typical feature (NAGL, 1970). The human X-chromatin has not been reported to demonstrate anything similar.

The frequency of drumsticks in polymorph nucleated leucocytes in blood smears from normal female persons is particularly low (0.2–2%). The heterochromatic X-chromosome forms obviously an extra lobule infrequently. The degree of nuclear segmentation and several other factors influence this frequency (see TOLKSDORF, 1974).

Genetic Activity

LYON (1961) has shown that in the mouse one of the two X-chromosomes is indeed genetically inactive. The inactivation takes place in early embryonic life. Which one of the two X-chromosomes in a particular cell (the paternally or maternally derived X-chromosome) is inactivated, seems to be a random process. Once the inactivation occurs it is fixed in the cell's heredity. As a consequence females are mosaics with one part of their cells having the paternal X inactivated and the remainder having the maternal X inactivated.

A random inactivation of one or the other X-chromosome was also demonstrated in somatic cells of man for certain X-linked genes (see LYON, 1968; GARTLER et al., 1972; PASSARGE and FRIES, 1973). The cytological evidence that the region of the X-chromatin does not (or only to a low degree) transcribe was provided by COMINGS (1966) by means of ^{3}H-uridine labelling. The genetic inactivity is, however, not directly connected with a microscopically visible heteropycnosis. In logarithmically growing fibroblast cultures of female tissues, only about half of the cell nuclei contain a distinct X-chromatin body. In tissues stemming from a person heterozygous for the variants A and B of the X-linked glucose-6-phosphate dehydrogenase gene both variants are expressed (see GARTLER et al., 1972). If, however, a clone is obtained consisting of cells with the B-variant inactivated, the A-variant is active in all cells although half of them show no X-chromatin. This means that also in cells, where the heterochromatic X is not condensed it can still be inactivated (COMINGS, 1966b). Similar results were obtained from cells with multiple X-chromosomes (THERKELSEN and PETERSEN, 1967).

It has long been known that in early human embryos an X-chromatin cannot be seen with certainty (PARK, 1957). In other mammals it has been clearly demonstrated that X-chromatin formation takes place around the period of implantation (e.g. MELANDER, 1962). Thus the occurrence of heteropycnosis goes approximately parallel with genetic inactivation. It is, however, not quite clear at exactly which point of time the inactivation of one X-chromosome takes place. From genetic studies of the size and distribution of mosaic patches of cells a stage around implantation seems to be the most probable one (see e.g. LINDER and GARTLER, 1965; SCHWARZACHER, 1969; PASSARGE and FRIES, 1973). In earlier cleavage stages of mammalian female embryos both X-chromosomes are active (HOPPE and WHITTEN, 1972; LYON, 1972). In oocytes both X-chromosomes behave euchromatically and isopycnotically (OHNO et al., 1962). EPSTEIN (1969, 1972) first demonstrated by estimating the amount of glucose-6-phosphate dehydrogenase (G6PD) and hypoxanthine guanine phosphoribosyl transferase (HGPRT) in XX and XO mice that in ovulated oocytes of mice both X-chromosomes are active. For human oocytes of mature ovaries as well as from a 16 week old fetus the activity of both X-chromosomes was shown beyond doubt by GARTLER et al. (1972) and GARTLER et al. (1973).

Late DNA Replication

In Chapter V it has been stated that not all chromosome regions replicate DNA synchronously during the S-period. Certain regions start DNA replication somewhat later and finish it quite distinctly later than the rest. Late DNA-replication has turned out to be a phenomenon connected with heterochromatic chromosomes and chromosome regions. It occurs in facultative as well as in constitutive heterochromatin. Indeed, the first demonstration of late DNA-replication was given by LIMA DE FARIA (1959) for the sex chromosomes of the grasshopper. In mammals it was first shown by TAYLOR (1960) in the Chinese hamster, that one of the two X-chromosomes in females is particularly late DNA replicating. In female human cells this was shown by GRUMBACH *et al.* (1963, see Figs. 39 and 41).

Many studies (see e.g. GIANNELLI, 1970) have made it clear that the genetically inactive X-chromosome is the one which exhibits late DNA replication. Late DNA-replication is obviously a constant feature of heterochromatin. As has been shown above, heteropycnosis (strong condensation in interphase) is not. HEIL (1970) found that in human female fibroblast cultures all cells had a late DNA replicating X-chromosome although only about 50% of the cells revealed an X-chromatin body.

X-Chromosome Abnormalities

In cases of abnormal numbers of X-chromosomes, the inactivation and hence the presence of heterochromatic X-chromosomes follows a simple rule: One X-chromosome per diploid chromosome set is always active and euchromatic, whereas all surplus X-chromosomes are heterochromatic, and inactivated (see e.g. HAMERTON, 1971). In cases with structural abnormalities of one X-chromosome the conditions may be more complicated. Many observations exist indicating that in such cases the normal X-chromosome is the active and euchromatic one. The abnormal X-chromosome is usually late DNA replicating (see HAMERTON, 1971). Also the size of the X-chromatin has been reported to correspond to the structurally abnormal X-chromosome (KLINGER *et al.*, 1965). The pathological symptoms of patients may be explained by a selection of cells with the active normal X against those with the active abnormal X. The latter cells may well not be viable for long periods, and the loss of cells, as well as possibly and occasional genetic imbalance during embryonic development could cause the malformation. In cases of X-autosomal translocations, a position effect can be demonstrated (CATTANACH and ISAACSON, 1961; HAMERTON, 1971).

Some recent observations in translocated X-chromosomes and other anomalies of the X-chromosomes have brought new hypotheses concerning the mechanism of inactivation. COHEN *et al.* (1972) and SUMMIT *et al.* (1973) discuss the possibility that the X-inactivation is only possible if one or more regions of the X-chromosomes are intact. Such a region would be a hypothetical X-inactivation center. THERMAN *et al.* (1974) suggested on the basis of the morphology of the X-chromatin in cases with abnormal X-chromosomes that a center for the X-chromatin condensation may be situated on the proximal part of the long arm of the X-chromosome.

VIII. Heterochromatin

1. Introduction

In the previous Chapters heterochromatin was frequently mentioned and some of its aspects were discussed. It may now be useful to give a short review in order to coordinate the information on heterochromatin dispersed throughout this book.

2. Historical Remarks and Nomenclature

The term heterochromatin was introduced by HEITZ (1928, 1929). He defined heterochromatin as chromosomal material which does not decondense in telophase as do the rest of the chromosomes. Therefore, heterochromatin stays more condensed during interphase, and in early prophase it appears already more condensed ("precocious condensation"). HEITZ named this property "heteropycnosis". The term "heterochromatin" was derived from MONTGOMERY's (1906) term "heterochromosomes" for the deeply stained sex chromosomes in spermatocytes, which were first observed by HENKING (1891) in insects and designated with the symbol "X". The expression "heteropycnosis" had already been used by GUTHERZ (1907) for the state of different condensation of the sex chromosomes during spermatogenesis. ÖSTERGREEN (1950) introduced the term "isopycnosis" for normally behaving chromosomes, and at present it is common to use the term "euchromatin" for the non heterochromatic chromosomal material. Another term sometimes used to describe the different behaviour in the condensation-decondensation cycle is "allocyclic" (WENRICH, 1916; DARLINGTON and LACOUR, 1940).

HEITZ (1935) stated explicitly that heteropycnosis is a property of heterochromatin which may change and he writes also: „*Die Ursache der Heteropyknose kann nur in dem betreffenden Chromosom selbst liegen. Das schließt nicht aus, daß in der Zelle bestimmte Bedingungen erfüllt sein müssen, damit sie in Erscheinung tritt*" ("The reason for heteropycnosis can only lie in the chromosome itself. This does not exclude the fulfilment of certain conditions within the cell in order that it may really be observed" HEITZ, 1935). For the sex chromosomes in gametogenesis such a change in behaviour was already known: JUNKER (1923) noted that in the stone fly *(Perla marginata)* in the ambisexual males (with XX-chromosomes, producing both sperms and oocytes), the two X-chromosomes are heteropycnotic in spermatocytes but not so in oocytes. Very extensive observations on the variable state of heteropycnosis of the X-chromosomes of the water spider *(Gerris lateralis)* were made by GEITLER (1939, 1953). Also in amphibians, heteropycnosis is very variable during development (BARIGOZZI, 1950). Another striking example of different states of heteropycnosis was discovered by SCHRADER (1929; see also HUGHES-SCHRADER, 1935, 1948): In coccids one complete chromosome set is heteropycnotic in diploid somatic cells of the male, whereas in females both sets are isopycnotic. In male gametes and in early embryonic cells also, both genomes are isopycnotic but one becomes heteropycnotic during development. All these examples demonstrated that in certain cases heteropycnosis is an unstable property of heterochromatin. It also became clear that since in all these cases only one homologous chromosome region becomes heteropycnotic, the reason for this cannot lie in a primary difference of the chromosome but rather in a difference of conditions. HUGHES-SCHRADER (1935) already noted therefore that heterochromatin may be facultative. In mammals one of the two X-chromosomes in females is the best known example of facultative heterochromatin.

On the other hand, many examples are known of heterochromatic chromosomes or chromosome regions whose state of heteropycnosis obviously does not change and which are present in both homologous chromosomes. The first example investigated by HEITZ *(Pellia endiviifolia)* as well as the heterochromatin in *Drosophila* belong to these. HEITZ (1933) described originally two types of heterochromatin (α and β) in *Drosophila virilis,* differing in their degree of condensation, but both types being stable and present on both homologues. These observations indicated that in certain types of heterochromatin this must rather be a property inherent to the chromosome.

This sort of heterochromatin is indeed different from the obvious facultative heterochromatin, and has been named "constitutive" heterochromatin. It has been shown, however, that constitutive heterochromatin also undergoes changes and is e.g. not present in all stages of development of *Drosophila* (COOPER, 1959). BROWN (1966) therefore defined that a heterochromatic segment of a chromosome is a region which is regularly and frequently observed to become heterochromatic.

The two types of heterochromatin are defined by BROWN (1966) in the following way:

Constitutive heterochromatin is regularly seen, in autosomes it is present on both homologous chromosomes and responds in the same way in both homologues during development.

Facultative heterochromatin may be observed only on one homologous chromosome. The two homologous chromosomes differ in that one becomes heterochromatic during development and the other remains euchromatic.

Of great importance was the demonstration by MULLER and PAINTER (1932) and by HEITZ (1933) that heterochromatic regions are genetically inactive. Another important discovery was that heterochromatic chromosomes are also allocyclic in respect to DNA-synthesis, being late replicating (LIMA DE FARIA, 1959). Other morphological particularities of heterochromatin have been found besides heteropycnosis, such as secondary constrictions (see WHITE, 1954), chromatid apposition (SCHMID, 1967), and increased frequency of breaks (HANNAH, 1951). Finally, enrichment of repetitive DNA (PARDUE and GALL, 1970; YUNIS and YASMINEH, 1970) and a differential stainability of constitutive heterochromatin have been demonstrated (LEVAN, 1946; ARRIGHI and HSU, 1971).

Currently the definition of facultative and constitutive heterochromatin as given by BROWN (c.f.) is widely accepted. In addition we know that constitutive heterochromatin is structurally a special sort of chromatin containing a high amount of highly repetitive DNA. Facultative heterochromatin must be considered as a special *state* of chromosomal material which is primarily not different from its homologous euchromatic material.

In general both the facultative as well as the constitutive heterochromatic regions have the following features in common:

1. Morphologic particularities (heteropycnosis, secondary constrictions, chromatid apposition);
2. Late DNA replication;
3. Genetic inactivity.

Constitutive heterochromatin is in addition further characterized by chemical differences (e.g. repetitive DNA) and a differential stainability by the C-band methods.

It should be mentioned, however, that not all these properties must be invariably found in any given case of heterochromatin. In some way they are, however, connected and dependent on each other. Furthermore it should be remembered that obviously different degrees of "heterochromatization" exist. HEITZ (1934) already described two types, named α and β heterochromatin. The two types differ in their degree of heteropycnosis, only the α-heterochromatin forming the deeply stained chromocenter. Both types belong to what we call today constitutive heterochromatin.

It has already been discussed in Chapter V that the G-bands may represent regions which are altered in the direction of heterochromatin. The definition of heterochromatin versus euchromatin is therefore based on a subjective judgement of the degree of the differences.

3. Constitutive Heterochromatin

Constitutive heterochromatin has been demonstrated so far in many different plant and animal species. Besides the earlier work on mosses (HEITZ, 1928), on the B-chromosomes in maize (for review see BATTAGLIA, 1964; HIMES, 1967), on *Drosophila* and other classical objects, the occurrence of constitutive heterochromatin was demonstrated using the banding techniques in a wide variety of organisms (e.g. VOSA, 1970; GALL *et al.,* 1971; HSU and ARRIGHI, 1971; BARIGOZZI and HALFER, 1972; NARDI *et al.,* 1973; CZAKER, 1975).

In the human genome, constitutive heterochromatin can be detected in almost all chromosomes, as has been described in Chapter III (Fig. 19). Constitutive heterochromatin is always present near the centromeric region. In several chromosomes also regions besides the centromere contain C-band material (notably chromosomes No. 1, 9, 16 and the Y, see Fig. 19).

Heteropycnosis and Related Features

Some of the constitutive heterochromatic regions correspond to secondary constrictions, the one on the Y-chromosome to a rather strongly contracted chromosome region. It is impossible to correlate all constitutive heterochromatic regions to heteropycnotic chromatin particles in interphase. Only the largest region, the distal part of the long arm of the Y-chromosome can be identified as a small dense chromatin particle. This is possible by a comparison of cell nuclei stained with quinacrine and by the Feulgen method. The other regions are presumably too small. They may well correspond to the numerous very small denser chromatin particles visible in the interphase nucleus.

Heteropycnosis, that means strong condensation in interphase, is, however, not a constant sign or property of constitutive heterochromatin. As has been described for the facultative heterochromatic X-chromosome in Chapter VII, the constitutive heterochromatic regions also can show a different degree of condensation. This has been demonstrated in more favorable species than man, with larger blocks of heterochromatin (e.g. SCHMID, 1967). A particularly well investigated species is the European field vole *(Microtus agrestis)* which has

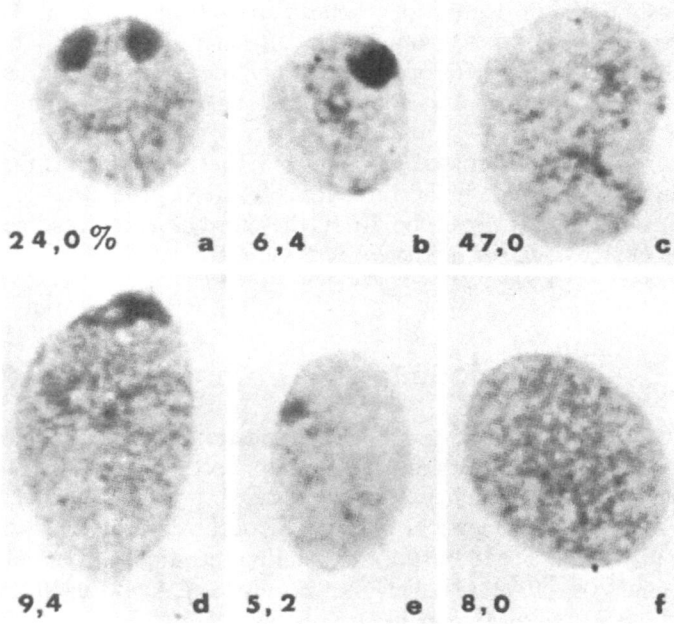

Fig. 94a–f. Cell nuclei of a kidney epithelial culture from a female *Microtus agrestis*. Examples for the different appearances and positions of the heterochromatic bodies formed by the large sex chromosomes. In (e) only the facultative heterochromatic part of the X-chromosome forms a body. In (f) no heteropycnotic bodies are seen. The %figures give the frequencies of types in a sample of 1000 cell nuclei. Feulgen stain. (From PERA, 1970)

very large sex chromosomes (MATTHEY, 1950) consisting to a great part of constitutive heterochromatin (WOLF *et al.,* 1965; SCHMID *et al.,* 1965). The two large blocks of constitutive heterochromatin form two large heteropycnotic bodies in interphase, but not in all types of cells (SCHMID, 1967; PERA, 1969). For instance kidney epithelial cells show them, but in cultured fibroblasts no condensed chromatin region is seen (Fig. 94). As has already been mentioned, the functional state of the cell may also influence the formation of the chromocenters. SCHNEIDER *et al.* (1973) showed that only inactive thyreoid cells of *Microtus agrestis* have condensed heteropycnotic bodies, whereas in stimulated cells they disappear. Here we should remember the statement by HEITZ (1935) cited above in which he says that expression of heteropycnosis may depend on certain conditions within the cell nucleus.

The differences in degree of heteropycnosis of constitutive heterochromatic regions do not seem to be connected with differences in genetic activity. For *Microtus agrestis* a lack of transcription of heterochromatin is also proved in cells in which heterochromatin is not heteropycnotic (SIEGER *et al.,* 1970).

In other cases, the lack of heteropycnosis may well indicate an activation of constitutive heterochromatin. In particular, the observations on early developmental stages are of importance (e.g. BARIGOZZI, 1950, for amphibians; COOPER, 1959, for *Drosophila*).

Direct observations on transcriptional activity and on DNA replication seem to be important in all these cases. The example of the facultative heterochromatic X-chromosome in female mammals shows us that in this instance heteropycnosis and late DNA replication go parallel with genetic inactivation (see e.g. LYON, 1968). In particular, the morphologic alterations of certain heterochromatic regions during gametogenesis in man (HUNGERFORD, 1971a; GAGNÈ *et al.,* 1974; STAHL *et al.,* 1975) are of great interest.

Some of the constitutive heterochromatic regions are seen as secondary constrictions. If it is assumed that these regions are more condensed in interphase, then they are characterized by a certain sluggishness in the condensation-decondensation cycle: they condense to a lesser degree during pro- and metaphase, and decondense to a lesser degree in telophase. That constitutive heterochromatic regions can be less condensed in mitotic chromosomes has been found in different mammalian species (see e.g. GROPP and NATARAJAN, 1972; SCHNEDL, 1971). In this connection the observation called "cold starvation" of heterochromatic regions should be remembered. DARLINGTON and LA COUR (1940) observed that heterochromatin is less stainable in plants grown at low temperatures. Many papers have been published dealing with this phenomenon. Recently LA COUR and WELLS (1974) reviewed the earlier observations and brought definitive proof that the chromosomes undergo a supercontraction due to the cold treatment, in which the heterochromatic segments are less condensed than euchromatic regions in metaphase. LEVAN (1946) who developed a differential stain for secondary constrictions suggested also that these may be characterized by a weaker condensation in metaphase. The findings that certain compounds which are built in into chromosomes of living cells (such as BrdU or the Bibenzimidazol-derivative 33258—Hoechst) cause the decondensation of heterochromatic regions point in the same direction (HILWIG and GROPP, 1973). Thus, from differential condensation properties two extreme sorts of constitutive heterochromatin can be described: 1. Regions which are more strongly condensed than euchromatin through mitosis and interphase. 2. Regions which are relatively strongly condensed in interphase but stay behind in mitotic condensation and hence are visible as secondary constrictions. In the human genome only the constitutive heterochromatin of the Y-chromosome belongs to the first type.

A further morphological characteristic of heterochromatic chromosomes is the so-called "chromatid apposition" (SCHMID, 1967). This is seen best in early metaphase, when the sister-chromatids containing longer blocks of constitutive heterochromatin lie very straight and close to each other. In the human genome this is again seen in the Y-chromosome. Facultative heterochromatin may also show chromatid apposition. Therefore it can sometimes be seen in one of the two X-chromosomes in female cells.

Late DNA-Replication

In all instances investigated so far, constitutive heterochromatic regions are late replicating (see e.g. COMINGS, 1972). A comparison of the late replication pattern in human chromosomes (see Figs. 39 and 19) with the C-banding shows this very clearly. With the autoradiographic method of demonstrating ^{3}H-thymidine, the end of the S-phase can be investigated quite easily (e.g. GERMAN, 1964b; SCHMID, 1963, see Fig. 39). The start of the S-phase is more difficult to investigate. Findings here are not so clear but there is no doubt that in man most chromosome regions which finish their replication late also start relatively late (e.g. CAVE, 1966; PRIEST et al., 1967). This has been studied in detail with the aid of the autoradiographic ^{3}H-thymidin method also in Microtus agrestis with its large blocks of constitutive heterochromatin (PERA and WOLF, 1967; PERA, 1968; SCHMID and LEPPERT, 1969). It could be shown that the constitutive heterochromatin needs about 4 hours for replication (Fig. 95). In Chapter V it was mentioned that the ^{3}H-thymidin autoradiographic method may not only indicate differences in the time course of DNA synthesis but also AT-richness. In the case of heterochromatin the ^{3}H-thymidin labelling is certainly reliable, since during early S-phase the heterochromatic chromocenters of Microtus agrestis cells are clearly unlabelled (Fig. 95). Furthermore, SPERLING and RAO (1974) demonstrated the late synthesis of heterochromatic chromosomes very convincingly by using the method of premature chromosome condensation after fusion of interphase cells with metaphases (Fig. 96).

Late replication of constitutive heterochromatin occurs also in those tissues in which no chromocenters are formed. This demonstrates that heteropycnosis per se is not related to late replication. Late replication has also been demonstrated for isolated DNA satellites (BOSCOTT and PRESCOTT, 1971).

In other than mammalian species the asynchrony in DNA-replication of constitutive heterochromatin is not always so clear. The relatively late finishing is found in all cases but the onset of DNA replication may occur at the same time as for euchromatin (see e.g. for Drosophila: KEYL and PELLING, 1963).

In Chapter V it has already been stated that the late replicating pattern in human chromosomes corresponds not only to the C-band pattern for the very late replicating regions, but also to a certain degree to the G-bands (Fig. 21). This again points in the direction that G-bands are regions of "slight heterochromatic" behaviour. In this connection the higher degree of condensation of the chromatin in the G-bands must be recalled (see Chapter V).

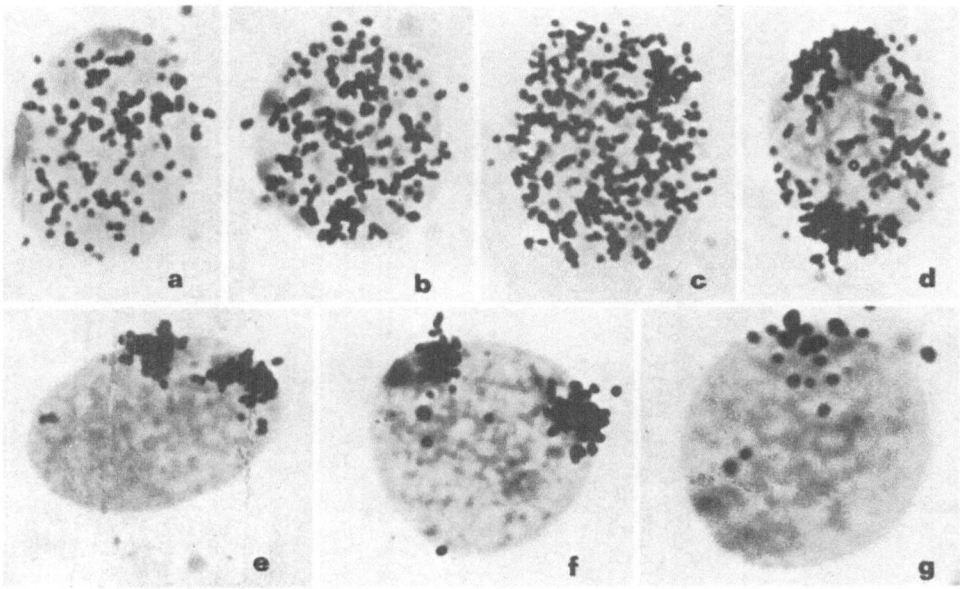

Fig. 95a. ³H-thymidin labelling pattern of female *Microtus agrestis* cell nuclei during the different phases (a–g) of the S-period. (From PERA, 1970)

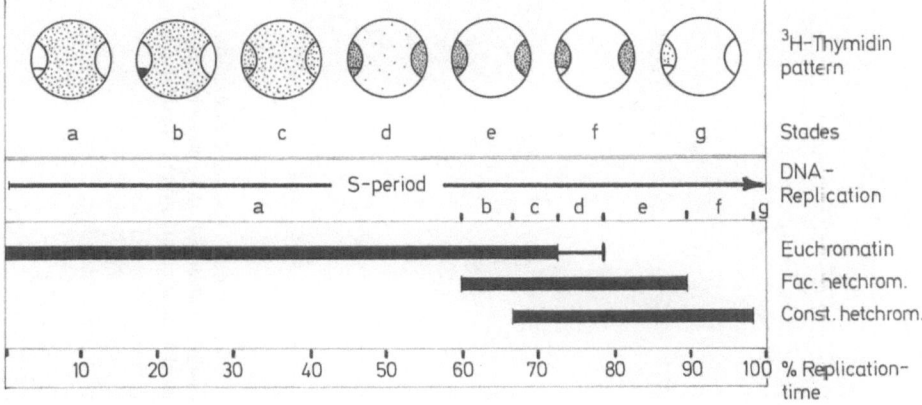

Fig. 95b. Schematic representation of the ³H-thymidin labelling pattern and the relative duration of the different phases (a–g) of the S-period. (From PERA, 1970)

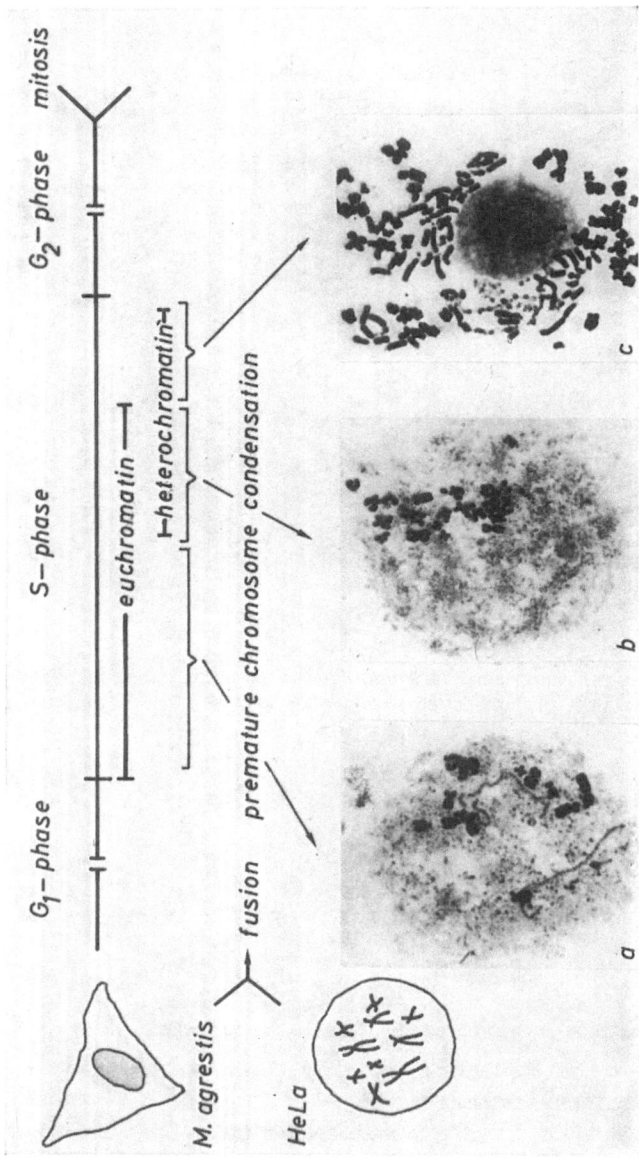

Fig. 96a–c. Prematurely condensed chromosomes of *Microtus agrestis* at early, middle and late S-phase. Fusion with mitotic HeLa cells. (a) Early S: The euchromatin is pulverized, the heterochromatin of the two X chromosomes is single stranded. (b) Mid S: The whole chromatin is pulverized. (c) Late S: The euchromatic chromosomes are double stranded, the late replicating heterochromatic regions are still synthetizing and hence appear pulverized. (From Sperling and Rao, 1974)

Chemical Composition

The differential staining with Giemsa after heat and alkaline treatment of constitutive heterochromatin (the C-band staining) indicates differences in chemical composition compared with euchromatin. Causal mechanisms of C-band staining have already been discussed in detail in Chapter V.

There seems to be no doubt that constitutive heterochromatin is rich in repetitive DNA. *In situ* hybridization of satellite DNA (PARDUE and GALL, 1970; JONES, 1970; SAUNDERS *et al.*, 1972; JONES *et al.*, 1973) has demonstrated that it is always concentrated in constitutive heterochromatic regions. Also acridine orange studies (DE LA CHAPELLE *et al.*, 1971 and 1973; BOBROW and MADAN, 1973) have indicated a high content of repetitive DNA. There are, however, some differences between different heterochromatic regions: the distal part of the long arm of the Y-chromosome seems to contain the highest amount of repetitive DNA among the constitutive heterochromatic bands in the human genome.

On the other hand not every region containing a higher amount of repetitive DNA is at the same time heterochromatic. The secondary constrictions in the short arms of the acrocentric chromosomes are known to represent the nucleolar organizers which contain repetitive DNA but reveal non of the characteristics of heterochromatin (OHNO, 1961; BIRNSTIEL *et al.*, 1968; LIMA DE FARIA *et al.*, 1969; HENDERSON *et al.*, 1972).

Some constitutive heterochromatic regions may also have a deviating CG/AT-relationship. The quinacrine staining of the heterochromatic region of the Y-chromosome indicates a relative enrichment of AT-bases (see Chapter V).

Another possibly important finding is that constitutive heterochromatin is relatively very rich in 5-methyl cytosin (MILLER *et al.*, 1974, see Chapter V).

Finally, many investigations have been carried out to demonstrate differences of protein components between heterochromatin and euchromatin. The main interest regarding differences in protein content was centred on histones since the demonstration by HUANG and BONNER (1962) that histones are able to suppress the transcriptional activity of DNA. It was first thought that either arginine-rich or lysine-rich histones were more important for this repressing action (see e.g. ALLFREY, 1969). Most of the earlier workers on histones, however, did not differentiate between facultative and constitutive heterochromatin, some even simply used condensed chromatin compared to less dense chromatin (e.g. ALLFREY *et al.*, 1963; FRENSTER, 1965). In a detailed study using clear examples of constitutive heterochromatin and euchromatin, COMINGS (1967) could not demonstrate striking differences in the composition of histones. SMART and BONNER (1971) found that no class of histones could be regarded as more important than the other with regard to the function of repressing template activity of DNA. Nevertheless the important findings of LITTAU *et al.* (1965) that only lysine-rich histones cause chromatin to condense should be remembered, particularly in connection with the observation of CLARK and FELSENFELD (1972) that AT-rich DNA is associated with lysine-rich histones.

Genetic Activity

A difference in genetic activity between heterochromatic and euchromatic chromosome regions was demonstrated for the first time in *Drosophila*. MULLER und PAINTER (1932) reported the Y- and half of the X-chromosome to be genetically "inert". HEITZ (1933) showed the genetically inert chromosome parts to be heterochromatic. These regions, however, had been shown to be not totally inactive: STERN (1927) had demonstrated fertility factors on the Y-chromosome. Later studies also showed that the other constitutive heterochromatic regions in *Drosophila* are not completely devoid of genes (e.g. BRIDGES, 1935; HANNAH, 1951; COOPER, 1959).

Direct cytologic evidence of genetic inactivity of constitutive heterochromatin in mammals came from autoradiographic studies demonstrating transcription activity. In cultured mouse L-cells HSU (1962) found only isopycnotic regions labelled after administering ^{3}H-uridine. More detailed studies on *Microtus agrestis* (SIEGER *et al.*, 1970) demonstrated very clearly that no transcription can be detected by the ^{3}H-uridine autoradiographic method in constitutive heterochro-

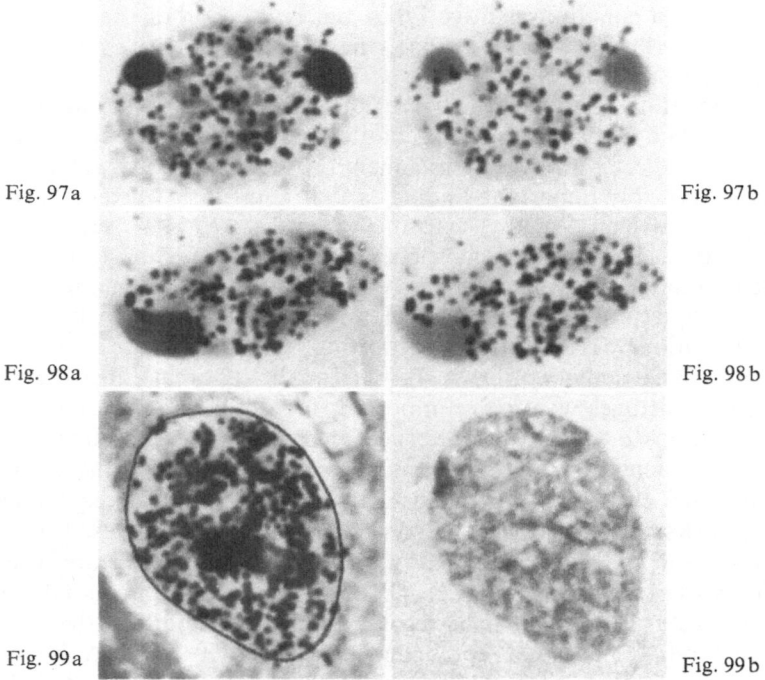

Fig. 97a   Fig. 97b

Fig. 98a   Fig. 98b

Fig. 99a   Fig. 99b

Figs. 97–99. Cell nuclei from *Microtus agrestis* after labelling with ^{3}H-Uridin. Regions with transcription activity labelled, regions without transcription activity unlabelled. Figs. 97 and 98: Brain cells (kept in suspension for 30 min after death). Staining with pararose anilin-methylgreen, photographed in green (a) and red (b) light. No label over the chromocenters. Fig. 99: Fibroblast in culture from a female animal. Only the facultative heterochromatic part of one X chromosome forms a heteropycnotic body. Lack of label over a much larger area, presumably the neighbouring constitutive heterochromatin. (According to SIEGER *et al.*, 1970)

matin. Of course, this method is not sensitive enough to detect a possible low transcription activity. The inactivity is not related to the state of condensation of constitutive heterochromatin. Cells showing no chromocenters in interphase behave like cells showing them (Figs. 97–99).

It should be pointed out here that genetic inactivity has also been very clearly demonstrated for examples of facultative heterochromatin, such as the heterochromatic genome in male coccids (NUR, 1966), or the heterochromatic X-chromosome in female mammalian cells (LYON, 1961, 1972). Thus there is no doubt that heterochromatic chromosome regions in general have no or only low genetic activity.

Activity in Meiotic Prophase

Recently, characteristic morphologic changes of constitutive heterochromatin have been reported in human meiotic prophase. GAGNÈ *et al.* (1971) described that the constitutive heterochromatin of chromosome No. 9 (visualized by the specific Giemsa-11 staining technique) forms small granules (so called "parameres") in the pachytene stage of the male first meiotic division. Similar granules were also observed by HUNGERFORD (1971) and HUNGERFORD *et al.* (1972). Further studies have revealed a similar change of the constitutive heterochromatin of chromosomes No. 9 in human fetal oocytes (GAGNÈ *et al.*, 1974). Moreover it was found that the constitutive heterochromatin if it is in this dispersed state is connected with nucleolar material. STAHL *et al.* (1974) observed in further

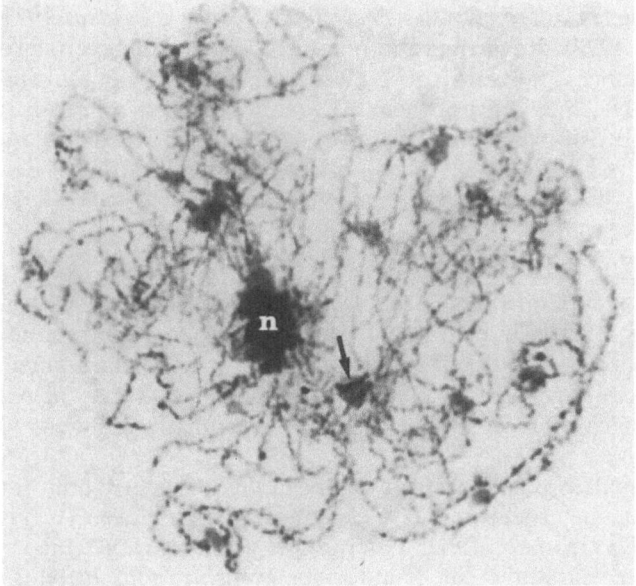

Fig. 100. Human oocyte in prophase I. Heterochromatic chromosome segments of chromosome No. 9 (arrow) associated with nucleolar material. Also other heterochromatic regions associated with nucleolar material. Courtesy of A. STAHL

studies that all constitutive heterochromatic regions presumably form small nucleoli during the first meiotic prophase in human oocytes (Fig. 100). If it is confirmed in future studies that constitutive heterochromatic regions are indeed transcribing RNA at this stage, a very important function of constitutive heterochromatin would be established.

Nucleolar Organizers

Besides the just mentioned possible activity of constitutive heterochromatic regions in meiotic cells, nucleoli in the vicinity of heterochromatin have been reported also in somatic cells. In human cultured fibroblasts and in human neurons a small nucleolus is frequently connected with the Y-chromosome in interphase (WYANDT and IORIO, 1973; IORIO and WYANDT, 1973). GAGNÉ et al. (1972) described the heterochromatic region of chromosome No. 9 being sometimes associated with a nucleolus-like body. HENDERSON et al. (1972) and EVANS et al. (1974) found no evidence for the presence of rRNA coding genes outside the already known nucleolar organizer regions at the acrocentric chromosomes, notably not in the Y-chromosome. These studies were made on chromosome preparations from lymphocytes in vitro by in situ hybridisation using ^{3}H labelled Xenopus laevis 18S and 28S RNA, or cRNA obtained from xenopus laevis rDNA. A possible "nucleolar structure" in the vicinity of the Y-chromosome may therefore indicate the collection of RNA other than 18S and 28S RNA. The same might be true for the reported "nucleoli" near the heterochromatic bands in spermatocytes and oocytes.

In other species constitutive heterochromatin may contain supposedly nucleolar organizing regions (e.g. LEVAN et al., 1962; NATARAJAN et al., 1971; NATARAJAN and GROPP, 1972). Regarding the relatively big size of heterochromatic blocks in these instances, it seems to be clear that nucleolar organizing regions can comprise only a small part of them. The main part of them seems to be inactive, as was described above. Nevertheless, the occurrence of nucleolar organizer within heterochromatic regions may play a role in the preservation of heterochromatin in evolution.

Position Effect

This was first demonstrated in Drosophila (MULLER, 1936; SCHULTZ, 1936 and 1941; LEWIS, 1950): If because of translocation an euchromatic chromosome region comes to be situated close to a heterochromatic segment, the former will also become genetically inactive. The inactivation may also conform with a morphological change of the original euchromatic segment which becomes heteropycnotic.

In mammals a position effect has been demonstrated only for cases where the facultative heterochromatic X-chromosome is involved (CATTANACH, 1961; OHNO and CATTANACH, 1962; CATTANACH and ISAACSON, 1965; EVANS et al., 1965). For recent studies on X-autosome translocations in man see COHEN et al. (1972), SUMMIT et al. (1973), THERMAN et al. (1974). No observations have been made so far on a position effect of constitutive heterochromatic translocations in mammals.

Lack of Chiasma Formation in Meiosis

The lack of crossing over between heterochromatic segments is of course evidenced by genetical studies demonstrating genetic inertness (MULLER and PAINTER, 1932; reviews in HANNAH, 1951, and COOPER, 1959). Cytologically, a lack of chiasma formation in the first meiotic division has been found in species with large blocks of heterochromatin (GROPP *et al.*, 1969, for the hedgehog; ZENZES and WOLF, 1971, for *Microtus agrestis*). The heterochromatic regions stay wide apart in

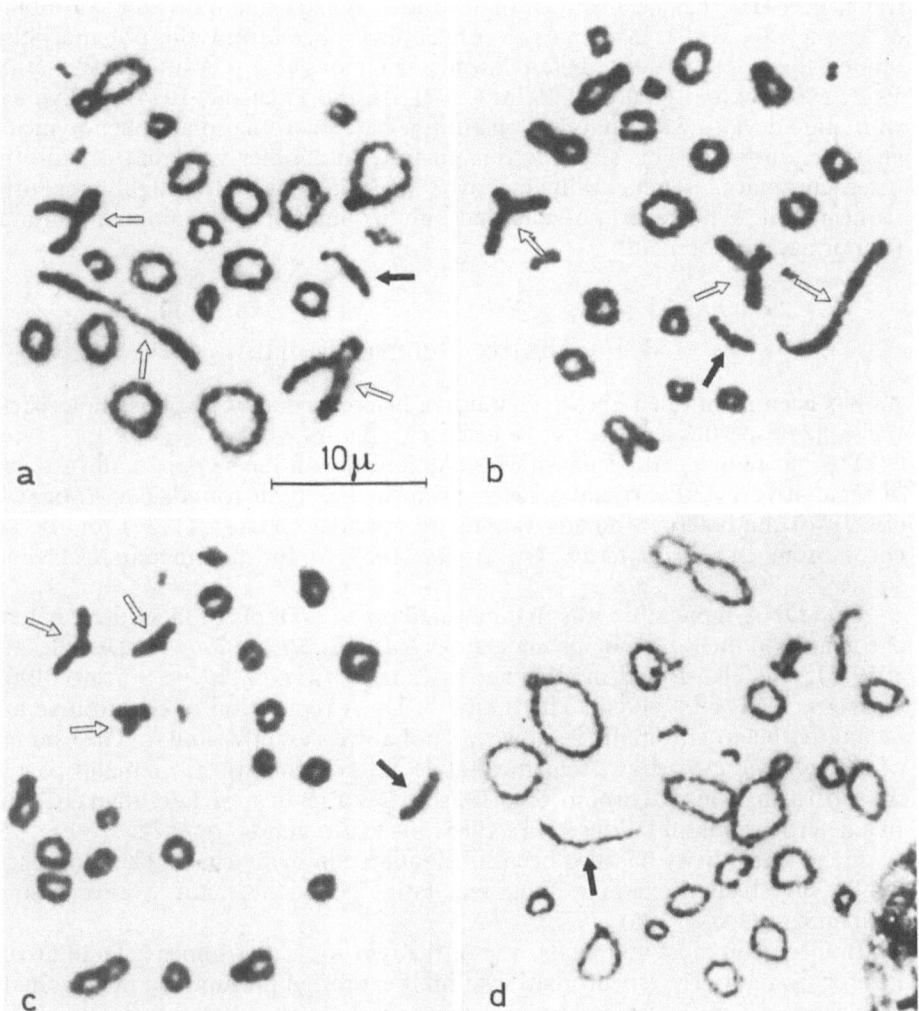

Fig. 101a–d. Meiotic figures from different male hedgehogs (late diakinesis and metaphase I). (a) and (b) *Erinaceus roumanicus*. (c) *Erinaceus europaeus*. (d) *Hemiechinus megalotis*. Dark arrow: The small X-Y bivalent. White arrows: Bivalents with spread chiasma free segments corresponding to heterochromatic segments. In *E. roum.* and *E. europ.* 3 large heterochromatic segments, in *H. megal*, no heterochromatin. (From GROPP *et al.*, 1969)

spermatocytes in the diplotene stage (Fig. 101). In human spermatocytes the heterochromatic regions can be visualized with the C-band method. Although they are very small and it is hardly possible to draw a negative conclusion from cytologic preparations it can be noted that also here the C-band regions have a tendency to stay apart.

Variability

In Chapter III it has been shown that a rather remarkable individual variability of the karyotype exists in man. At closer examination, one finds this polymorphism limited to the constitutive heterochromatic regions (e.g. CRAIG-HOLMES *et al.*, 1973; MÜLLER and KLINGER, 1974; GERAEDTS and PEARSON, 1974). Individuals with quite deviating amounts of constitutive heterochromatin are phenotypically absolutely normal. This seems to demonstrate in another way that constitutive heterochromatin is genetically inactive. Since it consists of highly repetitive sequences, it is, however, possible that only its amount is of minor importance for normal development.

4. Facultative Heterochromatin

As has been mentioned above, facultative heterochromatic regions show partly the same properties as constitutive heterochromatin.

The morphological peculiarities are practically all the same: In all instances of facultative heterochromatin heteropycnosis has been found, e.g. SCHRADER (1929) for the heterochromatic genome in coccids; GEITLER (1939) for the sex chromosome in *Gerris lateralis,* OHNO *et al.* (1959) for the mammalian X-chromosome.

Late DNA-replication was first detected on an example of facultative heterochromatin on the sex chromosomes of *Melanoplus differentialis* (LIMA DE FARIA, 1959). It was also found in all other cases (e.g. TAYLOR, 1960; SCHMID, 1963; GERMAN, 1964; NUR, 1966). The timing of DNA replication of constitutive and facultative heterochromatin is, however, not always exactly similar. The example of *Microtus agrestis* shows that facultative heterochromatin (in females part of the short arm in one X-chromosome) finishes DNA replication later than constitutive heterochromatin (WOLF *et al.*, 1965; SPERLING and RAO, 1974, see Fig. 95).

Genetic inactivity has also been substantiated in many cases. The most thoroughly investigated example is the facultative heterochromatic X-chromosome in mammals (LYON, 1961, 1972).

In man as in most mammals, one of the two X-chromosomes is facultatively heterochromatic. Heterochromatinization is expressed presumably in all somatic cells from a very early embryonic stage on. In oocytes and embryonic cells before implantation both X-chromosomes are euchromatic and genetically active.

The heterochromatic X-chromosome is heteropycnotic, forming the X-chromatin in interphase nuclei, late DNA replicating and genetically inactive. The details have already been discussed in Chapter VII under the heading "X-chromatin".

Here it should be added that the X-chromosome is also heteropycnotic in the male during the prophase of the first meiotic division. In this phase it forms, together with the Y-chromosome, the well known sex vesicle. Although the X and Y associate along only a small region (PEARSON and BOBROW, 1970) syneptinemal complexes can be demonstrated (FORD and WOOLLAM, 1966). Thus, exchange of genetic material may take place between the two sex chromosomes. Both sex chromosomes are somewhat less condensed in the sex vesicle than constitutive centromeric heterochromatin at the same time (COMINGS, 1972).

5. Function of Heterochromatin

Facultative Heterochromatin

The function seems to be clear in most cases. It is genetic inactivation of certain chromosome regions during certain stages of development or in certain tissues. In mammals, the inactivation of one X-chromosome in females serves as a dose compensation mechanism necessary for the balance between females with two X-chromosomes and males with only one X-chromosome.

Constitutive Heterochromatin

As has been already discussed in Capter V, constitutive heterochromatin contains highly repetitive DNA which is regarded as inactive or "trivial" DNA. OHNO (1972, 1974) supposes that trivial DNA has accumulated in the genomes during evolution by gene duplication. If a copy of a gene is created by duplication this redundant copy may be altered into a new functioning gene, or it will degenerate because of mutational changes which are not giving rise to a new useful gene. The latter process will occur much more frequently and hence a great amount of non-functioning trivial DNA will be accumulated.

Non-active or trivial DNA may play an important role as spacer between structural genes and may serve as a protection against mutagenic acting influences (OHNO, 1972). Since trivial DNA itself is inactive, its alteration would not do any harm.

The so called trivial DNA may amount to about 95% of all DNA in the mammalian genome. Constitutive heterochromatin comprises in most cases but a few percent (up to about 18% in some species). The role as spacer or absorber of mutagenic influences is presumably played mainly by the trivial DNA which is dispersed throughout the genome.

Constitutive heterochromatin on the other hand is a concentration of trivial DNA which is highly repetitive into a few larger chromosome regions or blocks. It can in some way be regarded as surplus-DNA, since species with large amounts of heterochromatin have more total DNA per genome than species with only few heterochromatin (SCHMID and LEPPERT, 1968). Since heterochromatin is, however, present in almost all species investigated so far, it may have some special function. Several speculations have been made in this respect, and some of these shall be briefly mentioned:

The position of constitutive heterochromatin near the centromere in so many different species has led to the suggestion that it may be important in some way for the kinetochore (JONES, 1970). It was also speculated that constitutive heterochromatin serves as a protection for kinetochores and nucleolus organizers (YUNIS and YASMINEH, 1971). No evidence in this direction, however, has been brought forward so far.

Another suggestion is that constitutive heterochromatin may have its function in attracting homologous chromosomes in meiosis. As is known from studies of the position of constitutive heterochromatin within the nucleus, a certain attraction does indeed take place (see Chapter IX), but this seems by far too weak to lead to a complete and exact chromosome pairing. Furthermore, in the mouse, all chromosomes possess the same satellite DNA, and therefore nonhomologous pairing should also occur.

The centromeric position of constitutive heterochromatin in many species may rather be a result of evolution. Chromosomal rearrangements in evolution seem to take place preferentially by involving the centromere (e.g. Robertsonian translocation). Small losses or imbalances of constitutive heterochromatin may not do much harm because of the high repetitiveness of DNA and its genetic inactivity.

Possible Genetic Acitivity of Constitutive Heterochromatin

All the functions of constitutive heterochromatin discussed above were based on the belief that it *never* possesses a special genetic activity. It has already been mentioned that constitutive heterochromatic chromosome regions are not heteropycnotic in all tissues and at all times during development. The change in heteropycnosis may well be a secondary phenomenon due to special conditions within the nucleus, as e.g. the varying appearance of the chromocenters in *Microtus agrestis* (SIEGER *et al.,* 1970; SCHNEIDER *et al.,* 1973). But in other instances, as for example in the early developmental stages of amphibians or in *Drosophila* (see BARIGOZZI, 1950; COOPER, 1959), the changes in heteropycnosis may also indicate changes in genetic activity. This supposition is supported by the observations that in developing amphibian embryos the DNA replicating pattern of heterochromatin changes (STAMBROOK and FLICKINGER, 1970). Of great importance in this connection are the observations of GAGNÉ *et al.* (1974) and STAHL *et al.* (1975) that in meiotic prophase constitutive heterochromatin may in some way well be genetically active. The evidence in this direction is by now only cytological: it has been shown in human spermatocytes and oocytes that in the direct vicinity of constitutive heterochromatin small nucleoli, containing RNA, are formed (Fig. 100). Further results have to be awaited, but it would be quite understandable, that constitutive heterochromatin could serve either as additional nucleolar organizer regions, or as other special RNA producing regions during the exceptional state of gametogenesis. Also in somatic cells the possibility exists that at least small parts of heterochromatic blocks contain nucleolar organizing (r-RNA-transcribing) regions (see e.g. LEVAN *et al.,* 1962; NATARA-JAN and GROPP, 1972; WYANDT and IORIO, 1973).

If we accept the hypothesis that constitutive heterochromatin is genetically active in meiotic prophase and perhaps also in early embryologic stages, some new aspects with regard to the current concept of heterochromatin are unavoidable. It is clear that constitutive and facultative heterochromatin are not as fundamentally different as was supposed. That means that although constitutive heterochromatin differs from facultative heterochromatin in that it consists of high amounts of repetitive DNA and perhaps special protein components, etc., this special constitution is not necessarily the *cause* of its being heterochromatic.

This would leave us with the concept that "heterochromatic" is indeed always a particular *state* of chromatin, and that even the mechanism of "heterochromatization" which is connected with inactivation may be the same in constitutive and facultative heterochromatin, in the case of constitutive heterochromatin affecting only a specially constituted chromosome region. Nevertheless, the differentiation between "constitutive" and "facultative" is important and should be used in future.

IX. The Position of Chromosomes within the Cell

1. Introduction

Much evidence exists for a non-random position of the chromosomes within the cell nucleus in interphase as well as within the "mitotic figure" in mitosis. Most earlier investigations were made on plants and lower animals, but during the last decade many observations have also accumulated showing a non-random position of chromosomes in mammalian and human cells.

Several types of order of the chromosomes within the cells can be observed. Four types will be discussed here:

1. A peripheral position of certain chromosomes or chromosome segments.
2. The association of chromosomes bearing nucleolar organizers.
3. The association of homologous chromosomes ("somatic pairing").
4. The gathering of chromosomes into complete chromosome sets ("genome separation").

The last two may be connected with each other and may both lead to a segregation of chromosomes ("somatic segregation"). Since genome separation may occur in the course of multipolar mitosis, this will also be considered in this chapter.

There is finally the question as to whether a particular position of the chromosomes is maintained during mitosis and interphase from cell cycle to cell cycle.

2. Peripheral Position

In man, the earliest observation of the peripheral position of a particular chromosome was made on the heterochromatic X-chromosome in female cells. The sexchromatin (see Chapter VII) is situated adjacent to the nuclear membran in almost all types of tissues (Figs. 89 and 90), a well known exception being nerve cells (e.g. BARR, 1966; MITTWOCH, 1967). Many observations indicate that condensed chromatin has in general a tendency to be situated peripherally in the vicinity of the nuclear membrane (see e.g. COMINGS, 1968). In cases of distinct chromocenters formed by constitutive heterochromatin they are, however, not always situated at the periphery (SCHMID, 1967; PERA, 1969; HSU and ARRIGHI, 1971, see Fig. 102). Thus, heterochromatin may have a certain tendency toward a peripheral position, but this is perhaps dependent on the functional state of the nucleus which in turn is connected with altered physicochemical conditions of the chromatin in general. If greater amounts of chromatin are strongly condensed (as is seen in many types of cells, a well known example being unactivated lymphocytes) the condensed chromatin is quite regularly situated near the nuclear membrane (e.g. GRUNDMANN and STEIN, 1961).

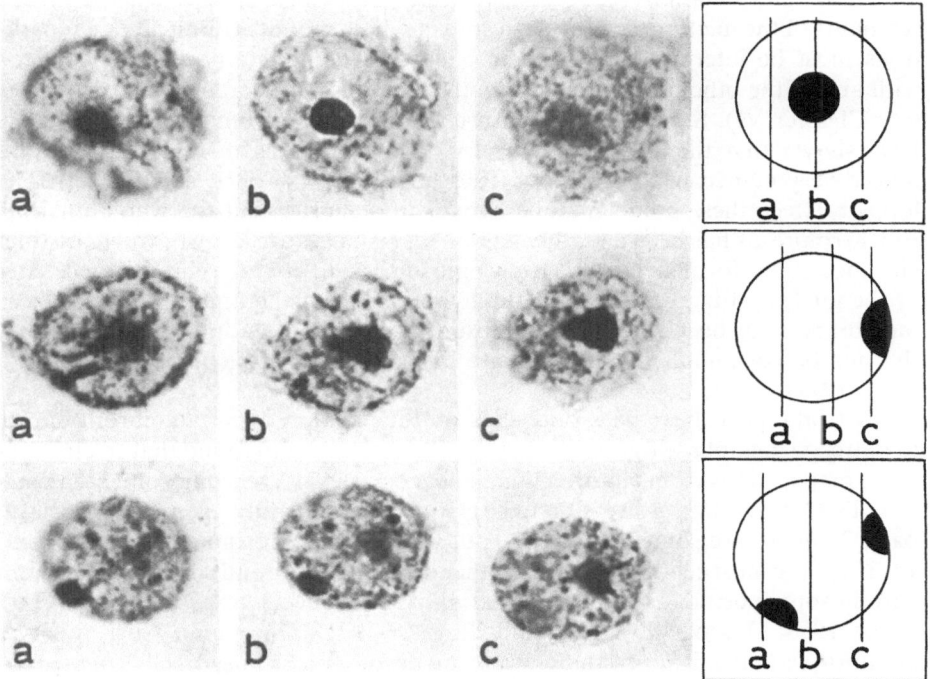

Fig. 102a–c. Cell nuclei of a brain cell suspension from a female *Microtus agrestis* photographed at different focus levels (a, b, c). *Above* nucleus with one central positioned chromocenter. *Middle row* nucleus with one chromocenter in peripheral position. *Bottom* nucleus with two peripheral chromocenters. Orcein stain, × 1600. (From PERA, 1969)

COMINGS (1968) discussed the possibility that in eukaryotic cells DNA replication may have its starting points at the nuclear membrane, as in parallel to the surface membrane of bacteria (JACOB *et al.*, 1966). COMINGS (1968) suggested that every chromosome may have several starting points of DNA replication which have to be connected with the nuclear membrane. In this way the chromosomes may be fixed in their position within the nucleus. Between these points the loops of the remaining chromosome regions extend into the inner parts of the nucleus. It has been shown, however, that no evidence can be found for the initiation of DNA synthesis at the nuclear membrane (e.g. O'BRIEN *et al.*, 1973; COMINGS and OKADA, 1973). The idea that each chromosome is attached to the nuclear membrane was thought to be supported by other observations: COMINGS and OKADA (1970) interpreted certain structures adhering to whole mount preparations of chromosomes in the electron microscope as parts (annuli) of the nuclear membrane. COMINGS (1972) proposed further, that since heterochromatin is more strongly condensed, the sites of attachment are also closer together and therefore heterochromatin appears to be situated near the nuclear membrane.

These interpretations must be critizised in several respects: The supposed remnants of the nuclear membrane in spread whole mount chromosome preparations must be interpreted as artefacts. No such remnants have been detected with any of the other electron microscopic methods on metaphase chromosomes (see Chapter VI). Studies on prematurely condensed chromosomes in S-phase have shown that the starting points for DNA synthesis in early S-phase are indeed very numerous and narrow (RAO and JOHNSON, 1971, see Fig. 83). The loops between these points may perhaps be too short to fill the whole nucleus. Furthermore, as has been already mentioned, instances are known, where distinct chromocenters formed by whole chromosomes are found right in the center of the nucleus (Fig. 102). Finally the hypothesis that the annuli of the nuclear membrane may be points of chromosome attachment is improbable because the number of annuli may vary quite considerably during interphase (MAUL et al., 1971).

Several reports have been published on the position of certain chromosomes in flattened metaphase figures. MILLER et al. (1963a and b) found that in chromosome preparations from cultured human leucocytes the Y-chromosome, the pairs No. 13, 21, 17 and 18 are situated relatively peripherally in metaphase, also pairs 3 and 16 are somewhat peripheral, whereas chromosome 1, 2, 19–20, and 14–15 are the most centrally located. Other investigators found somewhat different chromosome positions (e.g. BARTON et al., 1965; KOWARZYK et al., 1966; GALPERIN, 1968; OCKEY, 1969; HOO and CRAMER, 1971). One point which deserves attention is that all these studies were made on colchicinised cells, treated by a hypotonic solution and flattened by air drying or squashing. The colchicine may lead to a loss of the orderly arrangement because of the disorder of the mitotic spindle, and the hypotonic solution may lead to a further disarrangement. The differing results between various investigators may have technical reasons, but they could also represent true differences between individual cases. A certain tendency of some chromosomes to have a more peripheral position than others seems to be evident.

3. Association of Nucleolar Organizer Chromosomes

In the human karyotype 5 pairs of chromosomes have been shown without doubt to carry nucleolar organizer regions: chromosome No. 13, 14, 15, 21, and 22 (OHNO et al., 1961; BROSS and KRONE, 1972 and 1973; HENDERSON et al., 1972; EVANS et al., 1974). In all of them the nucleolar organizer regions correspond to a secondary constriction in the short arm between the centromeric region and the satellite. HARNDEN (1961) and FERGUSON-SMITH and HANDMARKER (1961) first observed that the satellite bearing chromosomes have a strong tendency to associate with their short arms in metaphase (Fig. 103). This phenomenon was further studied by several authors in normal and chromosomally abnormal cases (e.g. ZANG and BACK, 1968; ROSENKRANZ and FLECK, 1969). OHNO *et al.* (1961) proposed that this association is caused by the formation of big nucleoli by several nucleolar organizer chromosomes. The close association of nucleolar organizer regions may be preserved into mitosis and in that way produce the picture of "satellite association". IORIO and WYANDT (1973) and WYANDT and IORIO (1973) observed that in human cultured cells the Y-chromosome as well as the chromosome No. 9 are also sometimes in close connection with a nucleolus. They discuss the possibility that these chromosomes contain also nucleolar organizing regions. SPAETER (1975) indeed found in interphase nuclei a close association of the two Y-chromosomes in a 47XYY-case and also a certain association of the two homologous No. 9-chromosomes. This may, however, also be due to a general tendency of somatic pairing of homologous chromosomes, particularly if they contain larger blocks of constitutive heterochromatin.

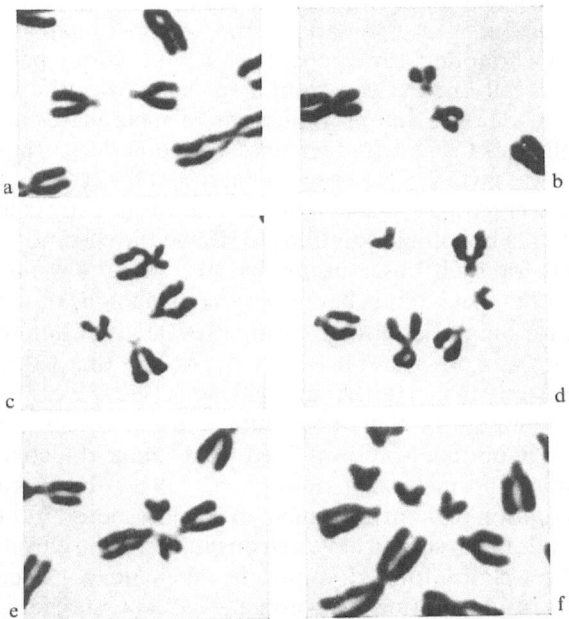

Fig. 103a–f. Association of human acrocentric chromosomes in different arrangements. (From PAS-SARGE, 1974)

4. Somatic Pairing

The association of homologous chromosomes is a regular feature of the first *meiotic* prophase in all species with a post reductional type of meiosis. It furnishes the possibility of an exchange of chromosomal material between homologues, the well known *meiotic crossing over*. It has been and still is a very important question to ascertain whether such an association of homologous chromosomes also occurs in *somatic* cells, because a somatic crossing over could then be possible. The genetic implications of a somatic crossing over are evident: cell clones of different genetic composition could arise within one individual.

Pairing of homologous chromosomes in somatic cells has been described in certain instances. The first observations were made in dipterans by STEVENS (1908) and METZ (1916). The well known giant chromosomes in the salivary gland cells of dipterans are paired polytene homologues (HEITZ and BAUER, 1932; BAUER, 1938). STERN (1936) provided the first genetical evidence, that somatic crossing over also takes place in *Drosophila*.

Somatic pairing has also been described in several plant species (see e.g. WAGENAAR, 1969).

In mammalian and human cells somatic pairing has never been demonstrated other than in exceptional cases, or only a tendency for an association of certain homologous chromosomes has been reported. The existence of somatic pairing in man is therefore still a much disputed problem.

Direct cytologic evidence of somatic pairing in a mammalian tissue was given by GIBSON (1970) who found a metaphase plate with closely paired homologous chromosomes in a cell line derived from a rat kangaroo (the Pt-K1-line). This observation was made once among thousands of metaphases and moreover on an aneuploid cell line. Convincing evidence for somatic pairing as a regularly occurring feature was given by HENEEN and NICHOLS (1972). They studied cultured cells of the Asian deer *(Muntiacus muntjak) in situ* without any colcemid treatment and without applying hypotonic solutions to spread the chromosomes. The muntjak is well suited for such observations because of its low number (2n=6 in females, and 2n=7 in males) and characteristic morphology of the chromosomes, situated radially forming a flat disc. A strong prevalence of homologous chromosomes to associate was observed. It is very interesting that COHEN et al. (1972), working independently from HENEEN and NICHOLS (1972) were not able to detect a pairing of homologues in the same animal, using colchicinised cells which were treated with hypotonic solution, fixed in acetic acid-alcohol and flattened by the usual air drying method. COHEN et al. (1972) think that the colcemid may lead to a disruption of the arrangement of chromosomes. If this interpretation is correct the spindle fibres should have a strong influence on chromosome position when the metaphase plate is formed. It must be taken into consideration, however, that other factors of preparation (hypotonic treatment, spreading etc.) may also disrupt the *in vivo* position of chromosomes. Moreover, many observations show that the position of chromosomes is kept fairly constant within the nucleus through the whole mitotic cell cycle.

The discrepancy between the two observations, made *in situ* and after spreading, is important for the evaluation of the observations on human material which have been done on routine spread chromosome preparations. It has already been noted that a non-random position of chromosomes has been found by several authors (e.g. MILLER et al., 1963a and b; BARTON et al., 1965; GALPERIN, 1968; OCKEY, 1969; HOO and CRAMER, 1971). Since some of these observations are consistent, the routine chromosome preparation may nevertheless represent in some respects the *in situ* position.

In human material, a certain tendency for a pairing of homologous chromosomes was found by BARTON *et al.* (1965) and SCHNEIDERMANN and SMITH (1962). The latter reported a significant association of pairs 3 and 16, whereas the other two pairs investigated (No. 1 and 2) showed a tendency which was, however, beneath the 5% level of statistical analysis.

The observations by GROPP and ODUNJO (1963) that homologous chromosomes quite often reveal morphological similarities in their configuration has also been interpreted as evidence for somatic pairing (Fig. 104).

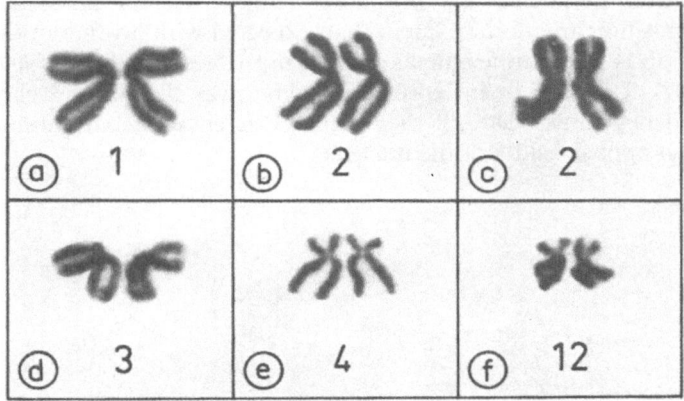

Fig. 104a–f. Concordance of the distortions of homologous chromosomes from metaphases of human lymphocyte cultures. (From GROPP and ODUNJO, 1963)

Another piece of evidence comes from observations of quadriradial configurations of two intimately associated chromosomes (GERMAN, 1964a). Such configurations are very rarely found in normal cells, but their frequency is increased in cell cultures from patients with Bloom's syndrome (GERMAN *et al.*, 1965) in which chromosome breaks are generally enhanced. GERMAN (1964a) reported that in 19 out of 21 observed quadriradial figures the two chromosomes were presumably homologous, while in 7 of them the chromosomes were identifiable without any doubt (Fig. 105). Quadriradial figures are the configurations expected in metaphase as a consequence of somatic crossing over (GERMAN, 1964).

Quadriradial formations are also observed after administration of agents which increase chromosome breaks. NOWELL (1964) and COHEN and SHAW (1964) reported quadriradial exchange figures in human peripheral blood cell cultures after treatment with mitomycin C. Quadriradials formed by two homologous chromosomes were always more frequent than those formed by nonhomologous chromosomes. RAO and NATARAJAN (1967) showed that several different agents which act as recombinogens lead to quadriradial figures of homologous chromosomes. More recently still other compounds have been found to induce quadriradials in human cells (e.g. GEBHART and BAUER, 1970). Thus, it seems to be clear that association of homologous chromosomes is not caused by a particular agent, but that different agents inducing chromosome breaks and recombination induce quadriradials. Therefore, quadriradials of homologous chromosomes are formed more frequently because homologous chromosomes are more often associated than non homologous ones.

Interphase nuclei have also been searched for signs of a somatic association or pairing of chromosomes. This is of course only possible for chromosomes which can be recognized in interphase. For instance, the heterochromatic sex chromosomes of *Microtus agrestis* form large chromocenters in interphase. Their position within the nucleus is non-random. Both extremes, an association and a vis à vis-position are more frequent than expected with a random distribution. This situation is found in female as well as in male cells (PERA, 1969, see Figs. 106 and 107). This may be interpreted in such a way that the sex chromosomes have a tendency to associate, if they come close enough, but cannot associate if they lie on opposite sides of the nucleus.

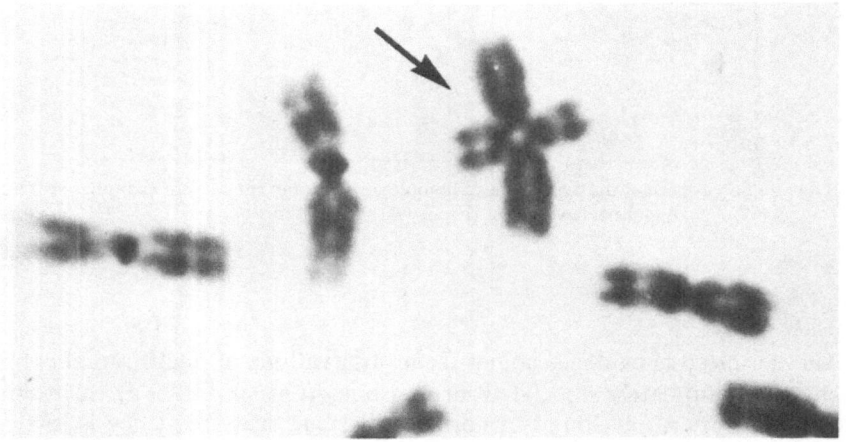

Fig. 105. Quadriradial figure formed by the two homologous chromosomes No. 6 from a human lymphocyte culture. G-banding stain. Courtesy of J. GERMAN

In human cells the special Giemsa-11 staining of the centromeric region of chromosome No. 9 is also seen in interphase as a purple dot (BOBROW *et al.*, 1972; GAGNÈ and LABERGE, 1972). Another chromosome easily seen in interphase is the Y-chromosome when stained with quinacrine. SPAETER (1975) used both these interphase markers in a study of human cultured cells of normal persons and of persons with a 47XYY chromosome constitution. It was found that the chromosomes No. 9 have a slight tendency and the Y-chromosome a stronger tendency to associate (Fig. 108).

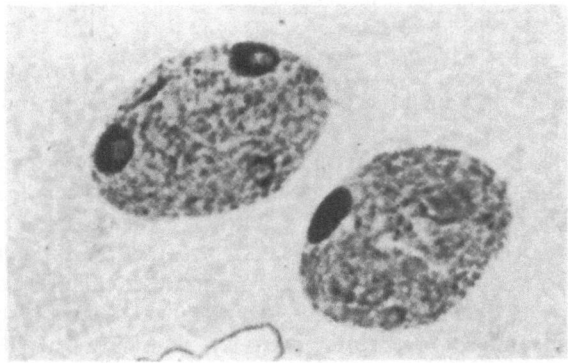

Fig. 106. Cell nuclei of a brain cell suspension from a male *Microtus agrestis*. One nucleus with two chromocenters the other with the two fused to one double sized chromocenter. Orcein stain. × 1 600. (From PERA, 1969)

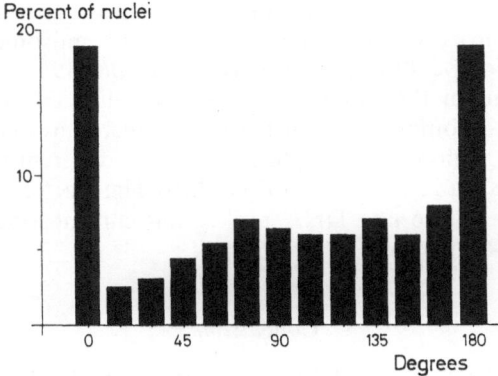

Fig. 107. Distribution of the angles between chromocenters and center of the nucleus. 0 degree corresponds to fused chromocenters, 180 degrees to opposite positioned chromocenters. Values from 200 nuclei of a kidney epithelial culture of a female *Microtus agrestis*. (From PERA, 1969)

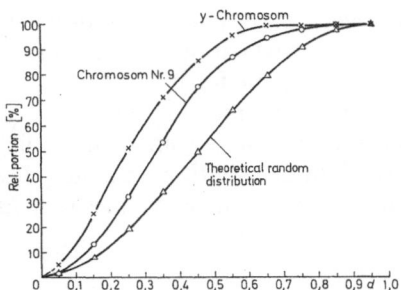

Fig. 108. Distribution curves of the distances between the two homologous chromosomes No. 9, and the two Y chromosomes (in a case with 47XYY chromosome constitution), compared with the theoretical courve for a random distribution. Measurements made on interphase nuclei from human lymphocyte cultures after "Giemsa-11" staining and quinacrine fluorescence staining respectively. (From SPAETER, 1975)

Mechanisms of Somatic Pairing

It is not known so far to which factor(s) the pairing of homologous chromosomes can be attributed.

In somatic cells the only relevant observations are the preference of heterochromatic chromosome regions to associate. In *Drosophila* it has long been known that all heterochromatic regions have a tendency to associate. They merge to form a big chromocenter (HEITZ, 1933). An association can also be observed between small non-homologous heterochromatin, so called "ectopic pairing" (SLIZYNSKI, 1945; HANNAH, 1951). The tendency to fuse is also demonstrated by the sex chromosomes of *Microtus agrestis* which consist to 75% of constitutive heterochromatin. In human cells the observations by SPAETER (1975) point in the same direction: the Y-chromosome with its larger region of constitutive heterochromatin has a stronger tendency to associate than the chromosome No. 9 with only a small heterochromatic region. The reason for the tendency of association of constitutive heterochromatin may lie in the repetitive DNA with many like sequences. In this connection it is of importance that no pairing of the facultative heterochromatic supernumary X-chromosomes was found in a case with the karyotype 49XXXXY (THORLEY *et al.,* 1967). However, as HSU (1974), pointed out, in the mouse, the same satellite DNA is present in the centromeres of all chromosomes, but the small heterochromatic particles corresponding to these regions in interphase are very differently distributed from tissue to tissue (HSU and ARRIGHI, 1971). Also HENEEN (1971) could not find a pairing tendency of the late DNA replicating chromosome segments in cell cultures of the rat-kangaroo.

Conclusions

True somatic pairing seems to be a very rare event in most mammalian and in human tissues. Also a somatic segregation of chromatids as a consequence of a somatic crossing over is presumably very infrequent. It may provide a "theoretical mechanism by which such findings as genetic differences in monozygotic twins, mosaicism of tissue or blood cell type in a single individual... might be explained" (GERMAN, 1964).

Somatic pairing is, however, not the only means by which a somatic segregation is caused. As will be shown in the next paragraph, genome separation and multipolar mitoses may be mainly responsible for it.

In addition to this seldom occurring close association of homologous pairs, one can observe a tendency of certain chromosome to lie more closely together than is expected with a random distribution. It is not clear what genetic consequences this non-random position of chromosomes may have.

5. Genome Separation

With this term the morphologic separation of complete chromosome sets or genomes is meant. In lower animal species and in plants such a separation is sometimes seen. One already mentioned example is the mealy bug (coccid) in whose somatic cells the paternally derived genome is heterochromatic and separated within the nucleus from the euchromatic maternal genome (HUGHES-SCHRADER, 1948).

In mammalian cells, the first indications for a genome separation were observed by GLÄSS (1957). He found that in mitosis of regenerating rat liver cells, the chromosomes often form groups which come in their number close to complete genomes. Most of the cells in which such separations were observed were polyploid. The techniques for visualizing chromosomes were not quite adequate for such observations at that time. Nevertheless some of the mitotic figures published by GLÄSS (1957) show a convincing division into distinct chromosome groups. However, polyploid and multinucleated cells in the liver arise presumably quite frequently by cell fusion (see PERA, 1970). Mitotic figures of fused cells very often show the chromosomes of the fused cells still in separated groups (HARRIS et al., 1966).

That a genome separation has occurred is indicated when cells having a ploidy differing from the original cell population are found. Such cells may arise by multipolar mitosis with a distribution of chromosomes in complete genomes. In cell cultures of *Microtus agrestis* haploid and triploid cells could be demonstrated by PERA and RAINER (1973) and similar observations were made in *Rhesus* kidney cell cultures by RIZZONI et al. (1974). In human leukocyte cultures PAWLOWITZKY and CENANI (1967) have reported a triploid cell.

6. Multipolar Mitosis

The first multipolar mitoses were described by MAYZEL (1875) in animals and by STRASBURGER (1880) in plants. BOVERI (1888) believed that the distribution of chromosomes in multipolar mitoses is a matter of chance. Since then multipolar mitosis was thought to be a pathological event (see e.g. STERN, 1958). But as early as 1908 BALTZER (1908) observed normal developing larvae after tripolar mitoses of dispermic triploid sea-urchin eggs. This indicated that multipolar mitoses may produce daughter nuclei which contain a reduced number of chromosomes ("somatic reduction"), but, at the same time, complete genomes. The distribution of chromosomes in complete genomes by multipolar mitoses was also considered by KOSTANECKI (1911) and RUTISHAUSER (1963). OHNO (1966) reviewed the possibilities of somatic segregation and discussed how this can be established by a quadripolar mitosis of a tetraploid cell (Fig. 109).

Direct cytological observations of multipolar mitoses with a distribution of the chromosomes in whole chromosome sets were made in cultured cells of *Microtus agrestis* (PERA and SCHWARZACHER, 1969; PERA, 1970). In this material the genomes are marked by the large, heterochromatic sex chromosomes. Their distribution in multipolar anaphases, in the daughter nuclei, and DNA-measurements of the daughter ana-telophase figures as well as the daughter nuclei established almost beyond doubt the genome-wise chromosome distribution. Multipolar mitoses were predominantly found in polyploid cells. All possible sorts of

genome distributions were observed, but tripolar mitoses of tetraploid cells result-
ing in 1 diploid and 2 triploid daughter nuclei were most frequent. Fig. 110
shows schematically the possible distributions (PERA, 1970). In Fig. 111 a tripolar
mitosis is shown with a genome distribution of 3:3:2, and in Fig. 112 three
daughter nuclei are shown indicating a genome distribution of 4:3:1. PERA and
RAINER (1973) were also able to find exact haploid and triploid mitoses in such
cultures. This indicates that such cells are even able to proliferate. RIZZONI
et al. (1972) could find similar chromosome distributions in multipolar mitoses
in cell cultures of *Rhesus* monkeys.

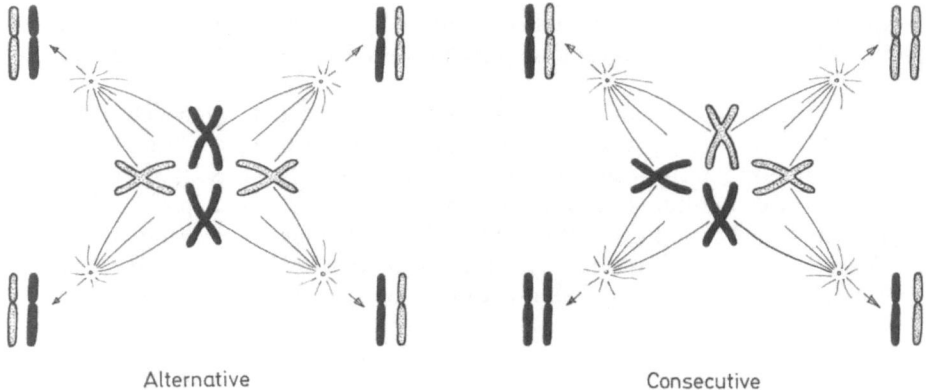

Alternative   Consecutive

Fig. 109. Schematic representation of somatic segregation by quadripolar mitosis of a tetraploid
cell. With a consecutive chromosome arrangement segregant daughter cells can appear. (After OHNO,
1966)

There seems to be little doubt that in pathological material, such as cancer
tissues or aneuploid cell lines, multipolar mitoses lead to an irregular chromosome
distribution (for lit. see PERA, 1970). This might be the case in normal tissues
also. The above cited observations show, however, that quite frequently multipolar
mitoses divide the mother nucleus into euploid daughter nuclei. The mechanism
by which this is done is still obscure. It has been found that an increase in
the relative occurrence of polyploid cells increases the number of multipolar
mitoses (SCHMID, 1966; HENEEN et al., 1970; PALITTI and RIZZONI, 1972). It
is, however, not clear whether differences exist when polyploid nuclei arise by
cell fusion or endomitosis. HENEEN (1970) working with rat-kangoroo cells showed
that multipolar spindles arise from multipolarly distributed centrioles. Sometimes
the multiple centrioles give rise to parallel spindles, sometimes radial symmetric
figures are formed.

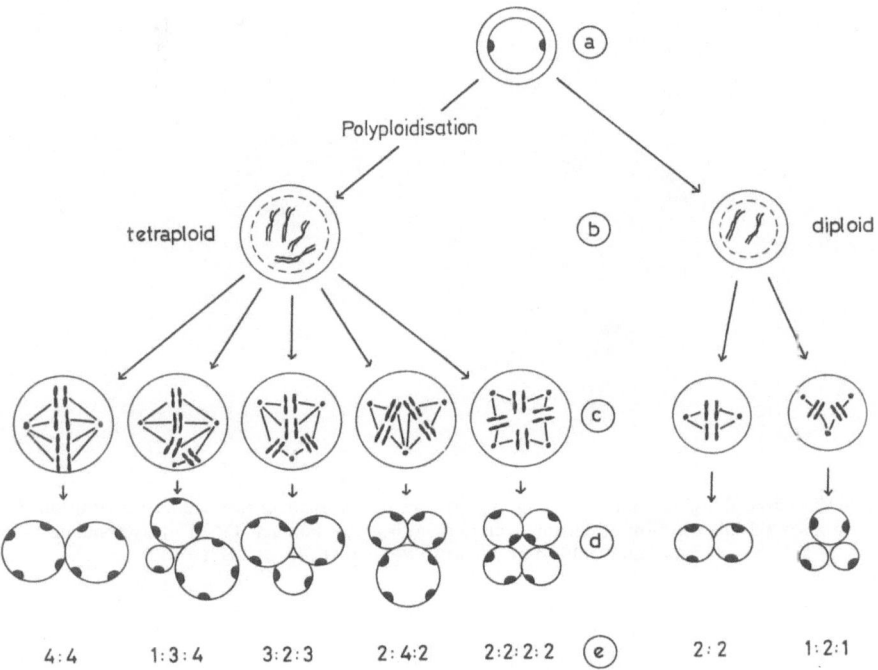

Fig. 110. Schematic representation of the distribution of chromosomes in various multipolar mitoses. The chromocenters in *d* indicate one haploid chromosome set. (From PERA, 1970)

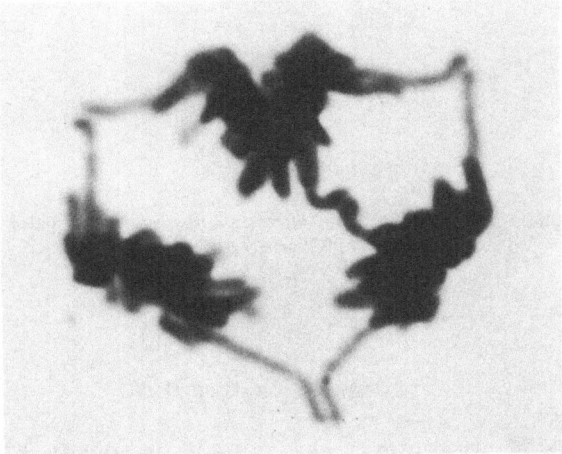

Fig. 111. Tripolar mitosis of a tetraploid cell from *Microtus agrestis*. The ploidy of each of the three daughter groups is indicated by the large sex chromosomes not yet quite separated. Chromosome distribution is 3:3:2. (From PERA and SCHWARZACHER, 1969)

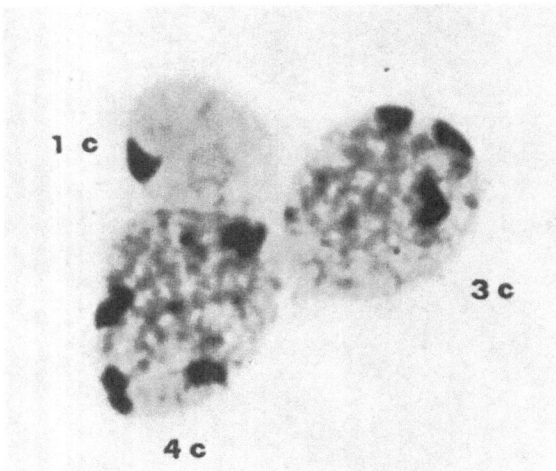

Fig. 112. Three daughter nuclei from a tripolar mitosis with a chromosome distribution 1:3:4, indicated by the distribution of chromocenters as well as by Feulgen DNA measurements. Kidney epithelial culture of *Microtus agrestis.* (From PERA, 1970)

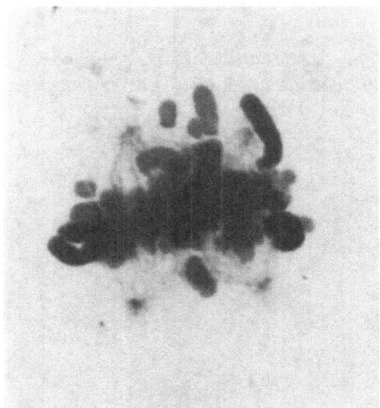

Fig. 113. Mitotic figure in a cell culture from *Microtus agrestis* with two parallel positioned mitotic spindles. Courtesy of F. PERA

7. Somatic Segregation

The evidence for a segregation of chromosomes in somatic cells of mammals, birds and fishes has been discussed by OHNO (1966, see also for earlier literature). More evidence was reported by SINHA (1967), TEPLITZ *et al.* (1968) and MARTIN and SPRAGUE (1969). The most convincing observations are those in cell hybrids by TEPLITZ *et al.* (1968). After fusion of cells from cattle and from mink, "tetra-ploid" hybrids containing a diploid set of each of the parental cells were observed. After a certain period of time "diploid" cells developed containing one haploid set of each of the original parental cells. Thus, reduction from "tetraploid" to "diploid" must have taken place in connection with a segregation of the chromosomes in whole sets. A similarly convincing observation was made by MARTIN and SPRAGUE (1969) in cultured human cells of patients carrying chromo-somal variants. After a period of spontaneous polyploidy diploid cells again occurred showing segregated pathological chromosomes.

In their studies of cultured *Microtus agrestis* cells, PERA and RAINER (1973, 1974) found, besides the occurrence of haploid and triploid cells, diploid cells with the sex chromosomes segregated, i.e. in male cultures cells with XX or YY constitution and in female cells with X_1X_1 or X_2X_2 constitution.

A somatic segregation could theoretically arise by a double nondisjunction (OHNO, 1966). But the rather frequent occurrence of chromosome segregation together with multipolar mitoses, as well as the findings of a chromosome distribu-tion in complete genomes with multipolar mitoses, point of course strongly to multipolar mitosis as the cause for somatic segregation. It is also possible that somatic pairing leads to a segregation of whole chromosomes, but somatic pairing seems to be—at least in cell cultures—a much rarer event than multipolar mitosis.

8. Constancy of Chromosome Position

Since we have seen that the position of the chromosomes within the cell nucleus or within the mitotic figure is not completely random, the question arises if a particular position is kept through interphase and mitosis.

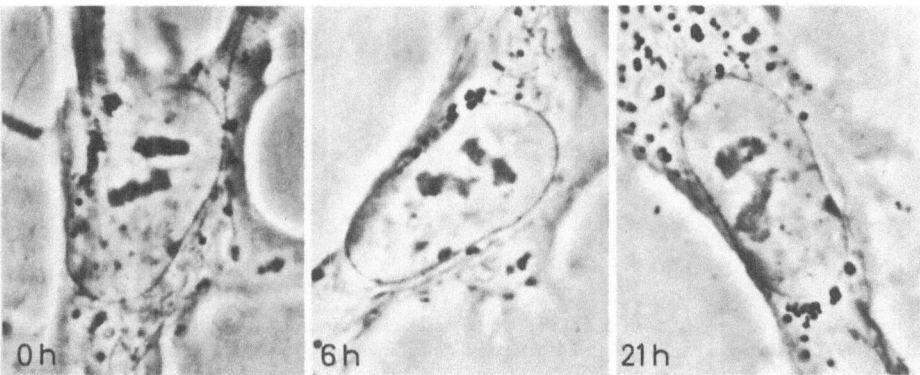

Fig. 114. A living cell from a human female fibroblast culture. Photographs taken at different times. The appearence of the X-chromatin (in the right to lower nuclear periphery) remains unchanged. Phase contrast. ×2000. (From SCHWARZACHER, 1963.) Courtesy of S. KARGER, publishers

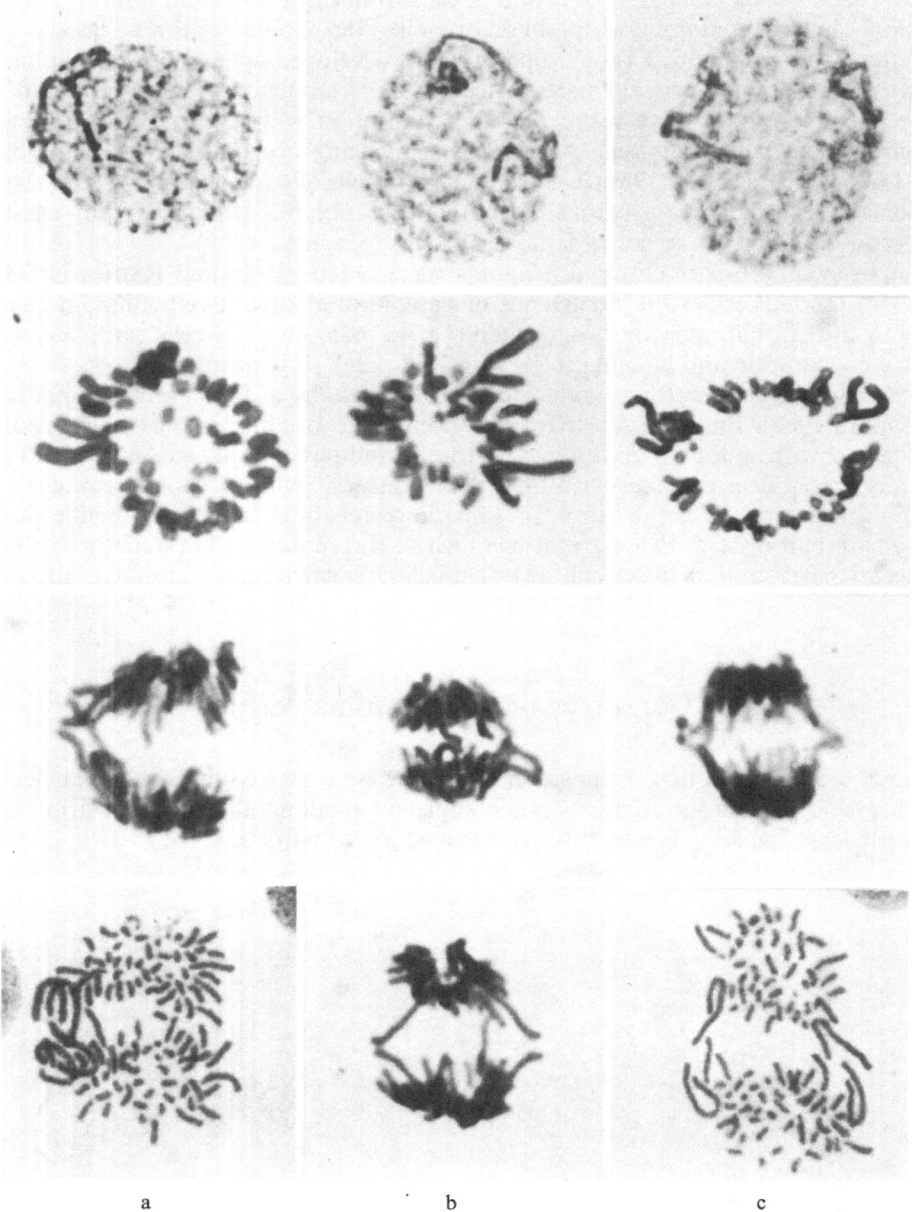

a   b   c

Fig. 115a–c. Three types of positions of the two large sex chromosomes in mitotic cells from *Microtus agrestis. First figure in each row:* The two sex chromosomes close together. *Middle figure:* Medium distance. *Third figure:* Opposite position. *Top row:* Prophases. *Second row:* Metaphases. *Third row:* Beginning anaphases. *Bottom row:* Late anaphases. Feulgen stain. × 1600. (From PERA and SCHWAR-ZACHER, 1970)

Direct observations relevant to this problem are scanty. In human tissues, the fact that the X-chromatin is relatively very frequently in a peripheral position speaks for a certain constancy of the position of the heterochromatic X-chromosome. In an observation of a living human female cell in culture, the relative position of the X-chromatin with respect to the nucleoli did not change for a period of 20 hours (SCHWARZACHER, 1963, see Fig. 114). In *Microtus agrestis* cells the position of the sex-chromosomes can be followed in both sexes because they form chromocenters in interphase and are by far the longest chromosomes easily recognizable also in mitosis. PERA and SCHWARZACHER (1969) observed that the position of these chromosomes is kept quite constant from interphase into mitosis and further into the daughter nuclei (Fig. 115). Sister nuclei coming from the same mother nucleus have therefore an identical pattern of chromocenter position (PERA and SCHWARZACHER, 1970, see Fig. 116). Thus, in those instances which allow the tracing of the position of a certain chromosome a constancy is observed. These instances are, however, at present restricted to heterochromatic chromosomes.

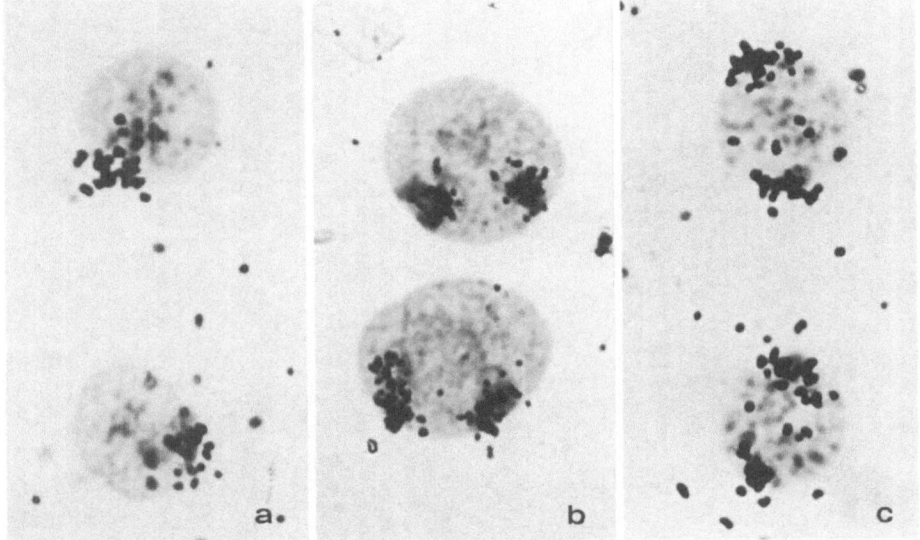

Fig. 116a–c. Sister nuclei with identical position of labelled Sex chromosomes. Kidney epithelial culture from a female *Microtus agrestis*. × 1 600. (From PERA and SCHWARZACHER, 1970)

In some non-mammalian examples a constant chromosome topography is also indicated, e.g. in turbellarians, the chromosomes have a certain non-random position in the sperms and in the first cleavage division (COSTELLO, 1970). A highly ordered arrangement of chromosomes is also found in nuclei of sperms from Salamanders (MACGREGOR and WALKER, 1973).

The mechanisms which may keep the chromosomes in their position are not yet clear. The possibility that chromosomes are attached to the nuclear membrane (COMINGS, 1968) is not proven, as has been discussed previously. But the observation by COMINGS (1968) that the spindle fibres attach to the chromosomes before the nuclear membrane has disintegrated at mitosis is very important. In this way the arrangement of the chromosomes, i.e. the position of their centromeres could be kept through mitosis.

Another mechanism discussed is the end-to end association of chromosomes seen in certain plants (WAGENAAR, 1969). Such an association of chromosomes has never been reported in mammalian and human cells.

It should be repeated in this connection that all the reports of interconnecting fibrils between chromosomes seen in the electron microscope (e.g. SHAW et al., 1972; BAHR et al., 1973) can be considered to be artefacts (see Chapter VI for a detailed discussion).

X. Summary and Conclusions

In this last chapter not so much a detailed summary will be presented but rather a concept of the structure of chromosomes in the light of the findings described in the foregoing chapters.

There can be no doubt that chromosomes are individual structural elements which are kept isolated from one another through the cell cycles and mitoses. No real evidence exists for any morphological interconnections between individual chromosomes. In somatic cells the only possible exception is the presumably very rare event of a pairing of homologous chromosomes and as a consequence, a somatic crossing over. Associations between chromosomes at sites with alike function, such as nucleolar organizers, is due rather to the gene product (the nucleolus) than to direct chromatin connections.

The most probable view is that a chromosome is composed of a single fibril containing one DNA-double helix molecule bound to proteins. This DNA-fibril is presumably seen in the electron microscope as a thread about 20–30 Å thick, which is folded into a thicker fibril of about 100 Å diameter, called the elementary fibril. It may have periodical thickenings caused by centrally located globules of histones. The elementary fibril is again folded in the form of small and irregular loops to form finally a chromatid. No regular foldings, such as a helical coiling can be observed. In interphase, the chromatids are closely packed and the loops protruding from each chromatid seem to intermingle. As soon as the chromatids condense, however, they separate from each other and can be seen as individual fine threads in the light microscope. This is especially well seen in preparations of prematurely condensed chromosomes. Such minimally condensed chromatids are not coiled but appear to be straight with only a few irregular bends. During the course of condensation this thin thread is laid into fairly regular coils (so-called "major coils"—although there are no "minor coils"), the number of coils decreasing but with increasing diameters. In this way the short and thick chromatids of metaphase chromosomes result.

During the S-period the DNA is replicated at many points. In prematurely condensed chromosomes of this period, the parts actually undergoing replication consist presumably only of the single-stranded DNA loops. These have been demonstrated only indirectly by ^{3}H-thymidine autoradiography bot not yet directly in the electron microscope. As soon as the DNA is replicated, each DNA-protein-fibril forms one chromatid which is morphologically separated from its sister chromatid. Only at the centromeric region do the two sister chromatids stick together, although no special structure connecting them can be seen.

Along the chromatids, differences exist which give them a banded appearance. The most prominent bands are regions of constitutive heterochromatin, which are in some cases (in man in the Y-chromosome) quite big, comprising the better part of a whole chromosome. These regions contain highly repetitive DNA, as well as relatively large amounts of 5-Methylcytosine, and are differently condensed than the rest of the chromosomes. Most of these regions are strongly condensed but some appear as secondary constrictions in metaphase chromosomes. The increased condensation lasts throughout the cell cycle. Larger blocks of heterochromatin may therefore form chromocenters in interphase. Heterochromatic regions are further characterized by their asynchronous DNA-synthesis and differences in their protein components. Constitutive heterochromatic regions are genetically inactive, or only very slightly active. Their functional significance seems to be unclear. Amongst possible functions, a specific genetic activity during gametogenesis, or some connection with the formation of the centromere may be mentioned. In the case of facultative heterochromatin, which are regions temporarily and inconstantly strongly condensed, asynchronously replicating, and genetically inactive, of course no primary structural differences to euchromatic regions exist. In mammals one of the two X-chromosomes in females is facultatively heterochromatic. This heterochromatisation serves as a dosage compensation in female cells.

Other regions of chromosomes can appear to be particularly less condensed in metaphase chromosomes. These are regions which contain ribosomal cistrons (for 18S and 28S RNA), the so-called nucleolar organizers. They contain also repetitive DNA.

Besides the constitutive heterochromatic regions proper there are numerous finer bands which can be made visible by a variety of methods (Q-, G-, R-bands). These fine bands presumably contain moderately repetitive DNA, which may be less repetitive than that in constitutive heterochromatic regions. Many bands are relatively rich in AT, but some are rich in GC. It has also been convincingly shown that they are regions of higher condensation, that means that the DNP-fibril is more tightly packed in the bands. The banding pattern is moreover similar to the late labelling pattern. Therefore at least several of the bands are also somewhat asynchronously replicating.

In slightly prematurely condensed chromosomes the bands are very small and numerous. Their number approaches that of the chromomeres observed in polytene giant chromosomes. It may therefore be justified to regard the fine bands as chromomeres. When the chromosomes contract in the course of mitosis, neighbouring bands which may form groups, fuse into the well known coarser bands of metaphase chromosomes. They are, however, independent of the major coils.

A chromosome can thus be visualized as a continuous DN-Protein fibril in which regions of stronger and slighter condensation alternate. The more tightly packed regions (the bands or chromomeres) contain a relatively higher amount of moderately repetitive DNA and are presumably somewhat later replicating than the more loosely packed regions (the interband zones). Other qualities such as high AT-content and differences in proteins may also add to the characterization of the bands. How far this segmentation of the chromosomes may be connected with the location of genes is open to speculation. Their number per genome is in the order of the presumed number of genes. Besides this organisation larger bands or blocks of constitutive heterochromatin are to be found in almost all chromosomes, and nucleolar organizing regions in some special chromosomes. Finally, the state of condensation or the degree of packing as well as the time course of DNA replication, and the genetic activity may be altered in quite large chromosome regions, depending on the functional requirement. In man this is the case in the facultative heterochromatic X-chromosome.

Reference

ABUELO, J.G., MOORE, D.E.: The human chromosome. Electron microscopic observations on chromatin fiber organization. J. Cell Biol. **41**, 73–90 (1969).

ALLFREY, V.G., FAULKNER, R., MIRSKY, A.E.: Acetylation and methylation of histones and their possible role in the regulation of RNA synthesis. Proc. nat. Acad. Sci. (Wash.) **51**, 786–794 (1964).

ALTMANN, H.W., GRUNDMANN, E.: Phasenkontrastmikroskopische Untersuchungen zur Vitalstruktur tierischer Zellkerne. Beitr. path. Anat. **115**, 313–347 (1955).

ANDERSON, T.F.: Techniques for the preservation of threedimensional structure in preparing specimens for the electron microscope. Trans. N.Y. Acad. Sci. **13**, 130 (1951).

ARRIGHI, F.E., HSU, T.C., SAUNDERS, P., SAUNDERS, G.F.: Localization of repetitive DNA in the chromosomes of *Microtus agrestis* by means of *in situ* hybridization. Chromosoma (Berl.) **32**, 224–236 (1970).

ARRIGHI, F.E., HSU, T.C.: Localization of heterochromatin in human chromosomes. Cytogenetics **10**, 81–86 (1971).

ARRIGHI, F.E., SAUNDERS, G.F.: The relationship between repetitious DNA and constitutive heterochromatin with special reference to man. In: Modern Aspects of Cytogenetics: Constitutive heterochromatin in man (R.A. PFEIFFER, ed.). Stuttgart-New York: Schattauer 1973.

ATKINS, L., BÖÖK, I.A., GUSTAVSON, K.H., HANNSON, O., HJELM, M.: A case of XXXXY sex chromosome anomaly with autoradiographic studies. Cytogenetics **2**, 208–232 (1963).

BADER, S., MILLER, O.J., MUKHERJEE, B.B.: Observations on chromosome duplication in cultured human leucocytes. Exp. Cell Res. **31**, 100–112 (1963).

BAHR, G.F.: Human chromosome fibres, considerations of DNA protein packing and of looping patterns. Exp. Cell Res. **62**, 39–49 (1970).

BAHR, G.F., GOLOMB, H.M.: Karyotyping of single chromosomes from dry mass determined by electron microscopy. Proc. nat. Acad. Sci. (Wash.) **68**, 726–730 (1971).

BAHR, G.F., GOLOMB, H.M.: Constancy of a 200 Å fiber in human chromatin and chromosomes. Chromosoma (Berl.) **46**, 247–254 (1974).

BAHR, G.F., MIKEL, U., ENGLER, W.F.: Correlates of chromosomal banding at the level of ultrastructure. In: Nobel Symposium **23**. Chromosome identification- technique and applications in biology and medicine (T. CASPERSSON, L. ZECH, eds.). London-New York: Academic Press 1973.

BAJER, A.: Change of length and volume of mitotic chromosomes in living cells. Hereditas (Lund) **45**, 579–596 (1959).

BAJER, A.: Subchromatid structure of chromosomes in the living state. Chromosoma (Berl.) **17**, 291–302 (1965).

BALDWIN, J.P., BOSELEY, P.G., BRADBURY, E.M., IBEL, K.: The subunit structure of the eukaryotic chromosome. Nature (Lond.) **253**, 245–249 (1975).

BALTZER, F.: Über mehrpolige Mitosen bei Seeigeleiern. Verh. phys.-med. Ges. Würzb. **39**, 291–330 (1908).

BARIGOZZI, C.: A general survey on heterochromatin. Port. Acta biol. Ser. A vol. R.B. GOLDSCHMIDT, 593–620 (1950).

BARIGOZZI, C., HALFER, C.: Constitutive heterochromatin in *Drosophila melanogaster*. Chromosomes today **3**, 8–11 (1962).

BARLOW, P., VOSA, C.G.: The Y-chromosome in human spermatozoa. Nature (Lond.) **226**, 961–962 (1970).

BARNICOT, N.A., HUXLEY, H.E.: The electron microscopy of unsectioned human chromosomes. Ann. hum. Genet. **25**, 253–258 (1961).

BARNICOT, N.A., HUXLEY, H.E.: Electron microscope observations on mitotic chromosomes. Quart. J. micr. Sci. **106**, 197–214 (1965).

BARR, M.L.: The significance of the sex chromatin. In: Internat. Rev. Cytol. vol. 19 (G.H. BOURNE, J.F. DANIELLI, eds.), p. 35–95. London-New York: Academic Press 1966.

BARR, M.L., BERTRAM, E.G.: A morphological distinction between neurones of the male and female, and the behavior of the nucleolar satellite during accelerated nucleoprotein synthesis. Nature (Lond.) 163, 676–677 (1949).

BARSKI, G., SORIEUL, S., CORNEFERT, F.: Production dans des cultures in vitro de deux souches cellulaires en association, de cellule de caractère "hybride". C.R. Acad. Sci. (Paris) 251, 1825–1827 (1960).

BARTON, D., DAVID, F.N., MERRINGTON, M.: The relative position of the chromosomes in the human cell in mitosis. Ann. hum. Genet. 29, 139–146 (1965).

BATTAGLIA, E.: Cytogenetics of B-chromosomes. Caryologia 17, 245–299 (1964).

BAUER, H.: Die polyploide Natur der Riesenchromosomen. Naturwissenschaften 26, 77 (1938).

BEERMANN, W.: Chromomeres and genes. In: Developmental studies on giant chromosomes (W. BEERMANN, ed.), p. 1–34. Berlin-Heidelberg-New York: Springer 1972.

BENDER, M.A., PRESCOTT, D.M.: DNA synthesis and mitosis in cultures of human peripheral leucocytes. Exp. Cell Res. 27, 221–229 (1962).

BENEDEN, E. VAN: Recherches sur la maturation de œuf et la fécondation. Arch. Biol. (Paris) 4, 265–638 (1883).

BERNHARD, W.: A new staining procedure for electron microscopical cytology. J. Ultrastruct. Res. 27, 250–265 (1969).

BIANCHI, N.O., BIANCHI, S.A.: DNA replication sequence of human chromosomes in blood cultures. Chromosoma (Berl.) 17, 273–290 (1965).

BICK, M.D., HUANG, H.L., THOMAS, C.A.: Stability and fine structure of eukaryotic DNA rings in formamide. J. molec. Biol. 77, 75–84 (1973).

BIRNSTIEL, M., SPEIRS, J., PURDOM, J., JONES, K., LOENING, U.E.: Properties and composition of the isolated ribosomal DNA satellite of Xenopus laevis. Nature (Lond.) 219, 454–463 (1968).

BISHOP, A., LEESE, M., BLANK, C.E.: The relative length and arm ratio of the human late-replicating X chromosome. J. med. Genet. 2, 107–111 (1965).

BOBROW, M., MADAN, K.: The effects of various banding procedures on human chromosomes, studied with acridine orange. Cytogenet. Cell Genet. 12, 145–156 (1973).

BOBROW, M., MADAN, K., PEARSON, P.L.: Staining of some specific regions of human chromosomes, particularly the secondary constriction of No. 9. Nature (Lond.) New Biol. 238, 122–124 (1972).

BOBROW, M., PEARSON, P.L., PIKE, M.C., EL-ALFI, O.S.: Length variation in the quinacrine-binding segment of human Y chromosomes of different sizes. Cytogenetics 10, 190–198 (1971).

BORDJADZE, V.K., PROKOFIEVA-BELGOVSKAYA, A.A.: Pachytene analysis of human acrocentric chromosomes. Cytogenetics 10, 38–49 (1971).

BOSTOCK, C.J., CHRISTIE, S.: Quinacrine fluorescence staining of chromosomes and its relationship to DNA base composition. Exp. Cell Res. 86, 157–161 (1974).

BOVERI, T.: Zellenstudien. Heft 2. Jena: G. Fischer 1888.

BOYES, J.W., NAYLOR, A.F.: Somatic chromosomes of higher Diptera. VI. Allosome-autosome length relations in Musca domestica. Canad. J. Zool. 40, 777–784 (1962).

BOYES, J.W., SLATIS, H.M.: Somatic chromosomes of higher Diptera. IV. A biometric study of chromosomes of Hylemya. Chromosoma (Berl.) 6, 479–488 (1954).

BRAMS, S., RIS, H.: On the structure of nucleohistone. J. molec. Biol. 55, 325–336 (1971).

BREG, W.R., ALLDERDICE, P.W., MILLER, D.A., MILLER, O.J.: Quinacrine fluorescence patterns and terminal DNA labelling of human C group chromosomes. Nature (Lond.) New Biol. 236, 76–78 (1972).

BREG, W.R., MILLER, O.J., MILLER, D.A., ALLDERDICE, P.W.: Distinctive fluorescence of quinacrine-labelled human G group chromosomes. Nature (Lond.) New Biol. 231, 276–277 (1971).

BRIDGES, C.B.: Salivary chromosome maps. J. Hered. 25, 60–64 (1935).

BRINKLEY, B.R., BRYAN, J.H.D.: The ultrastructure of meiotic prophase chromosomes as revealed by silver-aldehyde staining. J. Cell Biol. 23, 14A (1964).

BRINKLEY, B.R., STUBBLEFIELD, E.: Ultrastructure and interaction of the kinetochore and centriole in mitosis and meiosis. In: Advances in cell biology vol. 1 (D.M. PRESCOTT, L. GOLDSTEIN, E. MCCONKEY, eds.), p. 119–185. New York: Appleton-Gutury Crofts 1970.

BRITTEN, R.J., KOHNE, D.E.: Repeated sequences in DNA. Science 161, 529–540 (1968).

BROSS, K., KRONE, W.: On the number of ribosomal RNA genes in man. Humangenetik 14, 137–141 (1972).

BROSS, K., KRONE, W.: Ribosomal cistrons and acrocentric chromosomes in man. Humangenetik **18**, 71–75 (1973).

BROWN, S.W.: Heterochromatin. Science **151**, 417–425 (1966).

BURKHOLDER, G.D., OKADA, T.A., COMINGS, D.E.: Whole mount electron microscopy of metaphase-I chromosomes and microtubules from mouse oocytes. Exp. Cell Res. **75**, 497–511 (1972).

CALDERON, D., SCHNEDL, W.: A comparison between quinacrine fluorescence banding and ^{3}H-thymidine incorporation patterns in human chromosomes. Humangenetik **18**, 63–70 (1973).

CALLAN, H.G.: The nature of lampbrush chromosomes. Int. Rev. Cytol. **15**, 1–34 (1963).

CASPERSSON, T., FARBER, S., FOLEY, G.E., KUDYNOWSKI, J., MODEST, E.J., SIMONSSON, E., WAGH, U., ZECH, L.: Chemical differentiation along metaphase chromosomes. Exp. Cell Res. **49**, 218–222 (1968).

CASPERSSON, T., LOMAKKA, G., ZECH, L.: The 24 fluorescence patterns of the human metaphase chromosomes-distinguishing characters and variability. Hereditas (Lund) **67**, 89–102 (1971).

CASPERSSON, T., ZECH, L., JOHANSSON, C.: Differential binding of alkylating fluorochromes in human chromosomes. Exp. Cell Res. **60**, 315–319 (1970).

CASPERSSON, T., ZECH, L., JOHANSSON, C., MODEST, E.J.: Identification of human chromosomes by DNA-binding fluorescent agents. Chromosoma (Berl.) **30**, 215–277 (1970a).

CASPERSSON, T., ZECH, L., JOHANSSON, C., LINDSTEN, J., HULTÉN, M.: Fluorescent staining of hetero-pycnotic chromosome regions in human interphase nuclei. Exp. Cell Res. **61**, 472–474 (1970b).

CATTANACH, B.M.: A chemically-induced variegated-type position effect in the mouse. Z. Vererbungsl. **92**, 165–182 (1961).

CATTANACH, B.M., ISAACSON, J.H.: Genetic control over the inactivation of autosomal genes attached to the X-chromosome. Z. Vererbungsl. **96**, 313–323 (1965).

CAVE, M.: Reverse patterns of thymidine-H^3 incorporation in human chromosomes. Hereditas (Lund) **54**, 338–355 (1966).

CERVENKA, J., THORN, H.L., GORLIN, R.J.: Structural basis of banding pattern of human chromosomes. Cytogenet. Cell Genet. **12**, 81–86 (1973).

CHAMBERS, R.: The physical structure of protoplasm determined by micro-dissection and injection. In: General Cytology (E.V. COWDRY, ed.). Chicago: Univ. of Chicago Press 1924.

CHAPELLE, A. DE LA, SCHRÖDER, J., SELANDER, R.K.: Repetitious DNA in mammalian chromosomes. Hereditas (Lund) **69**, 149–153 (1971).

CHAPELLE, A. DE LA, SCHRÖDER, J., SELANDER, R.K.: Sequential denaturation and reassociation of chromosomal DNA in situ. In: Nobel Symposium **23**. Chromosome idenfication technique and applications in biology and medicine. (T. CASPERSSON, L. ZECH, eds.). London-New York: Academic Press 1973.

CHAPELLE, A. DE LA, SCHRÖDER, J., SELANDER, R.K., STENSTRAND, K.: Differences in DNA composition along mammalian metaphase chromosomes. Chromosoma (Berl.) **42**, 365–382 (1973).

Chicago Conference (1966): Standardization in human cytogenetics. Birth defects: Original article series, vol. II, No. 2. New York: The National Foundation 1966.

CHRISTENHUSS, R., BÜCHNER, TH., PFEIFFER, R.A.: Visualization of human somatic chromosomes by scanning electron microscopy. Nature (Lond.) **216**, 379–380 (1967).

CLARK, R.J., FELSENFELD, G.: Association of arginine-rich histones with G.C.-rich regions of DNA in chromatin. Nature (Lond.) New Biol. **240**, 226–229 (1972).

COHEN, M.M., ENIS, P., PFEIFER, C.G.: An investigation of somatic pairing in the muntjak (*Muntiacus muntjak*). Cytogenetics **11**, 145–152 (1972).

COHEN, M.M., LIN, C.C., SYBERT, V., ORECCHIO, E.J.: Two human X-autosome translocations identified by autoradiography and fluorescence. Amer. J. hum. Genet. **24**, 583-597 (1972).

COHEN, M.M., SHAW, M.W.: Effects of mitomycin C on human chromosomes. J. Cell Biol. **23**, 386–395 (1964).

COHEN, M.M., SHAW, M.W., MACCLUER, J.W.: Racial differences in the length of the human Y chromosome. Cytogenetics **5**, 34–52 (1966).

COLEMAN, J.R., MOSES, M.J.: DNA and the fine structure of synaptic chromosomes in the domestic rooster (*Gallus domesticus*). J. Cell Biol. **23**, 63–78 (1964).

COMINGS, D.E.: Uridine-5-H^3 radioautography of the human sex chromatin body. J. Cell Biol. **28**, 437–441 (1966a).

COMINGS, D.E.: The inactive X-chromosome. Lancet **1966 II**, 1137–1138.

COMINGS, D.E.: Histones of genetically active and inactive chromatin. J. Cell Biol. **35**, 699–708 (1967).

COMINGS, D.E.: The rational for an ordered arrangement of chromatin in the interphase nucleus. Amer. J. hum. Genet. **20**, 440–460 (1968).

COMINGS, D.E.: The structure and function of chromatin. In: Advanc. hum. Genet. **3**, 237–431 (1972).

COMINGS, D.E., AVELINO, E., OKADA, T.A., WYANDT, H.E.: The mechanisms of C- and G-banding of chromosomes. Exp. Cell Res. **77**, 469–493 (1973).

COMINGS, D.E., OKADA, T.A.: Do half-chromatids exist? Cytogenetics **9**, 450–459 (1970a).

COMINGS, D.E., OKADA, T.A.: Whole mount electron microscopy of the centromere region of metacentric and telocentric mammalian chromosomes. Cytogenetics **9**, 436–449 (1970b).

COMINGS, D.E., OKADA, T.A.: The association of chromatin fibres with the annulae of the nuclear membrane and its bearing on chromosome structure. Exp. Cell Res. **62**, 293–302 (1970c).

COMINGS, D.E., OKADA, T.A.: Fine structure of the kinetochore in the Indian Muntjak. Exp. Cell Res. **67**, 97–110 (1971).

COMINGS, D.E., OKADA, T.A.: Electron microscopy of chromosomes. In: Perspectives in cytogenetics. The next decade (S.W. WRIGHT, B.F. CRANDALL, L. BOYER, eds.), p. 223–250. Springfield/Ill.: Thomas 1972.

COMINGS, D.E., OKADA, T.A.: DNA replication and the nuclear membrane. J. molec. Biol. **75**, 609–618 (1973).

COOPER, H.L., HIRSCHHORN, K.: Enlarged satellites as a familial chromosome marker. Amer. J. hum. Genet. **14**, 107–124 (1962).

COOPER, K.W.: Cytogenetic analysis of major heterochromatic elements (especially Xh and Y) in *Drosophila melanogaster*, and the theory of "heterochromatin". Chromosoma (Berl.) **10**, 535–588 (1959).

COREY, M.J., BOYES, J.W., YERGENIAN, G.: Variations in relativ lengths of early and late replicating portions of Y chromosomes of the Chinese hamster (*Cricetulus griseus*). Cytogenetics **6**, 314–320 (1967).

CORNEO, G., GINELLI, E., POLLI, E.: Isolation of the complementary strands of a human satellite DNA. J. molec. Biol. **33**, 331–335 (1968).

CORNEO, G., GINELLI, E., POLLI, E.: Renaturation properties and localisation in heterochromatin of human satellite DNA's. Biochim. biophys. Acta (Amst.) **247**, 528–534 (1971).

CORNEO, G., GINELLI, E., ZARDI, L.: Satellite and repeated sequences in human DNA. In: Modern Aspects of Cytogenetics: Constitutive Heterochromatin in Man. Ed. R.A. PFEIFFER, Stuttgart-New York: Schattauer, 1973.

CORNEO, G., ZARDI, L., POLLI, E.: Elution of human satellite DNA's on a methylated albumin kieselgur chromatographic column: isolation of satellite DNA IV. Biochim. biophys. Acta (Amst.) **269**, 201–204 (1972).

COSTELLO, D.P.: Identical linear order of chromosomes in both gametes of the Acoel turbellarian *Polychoerus carmelensis*: A preliminary note. Proc. nat. Acad. Sci. (Wash.) **67**, 1951–1958 (1970).

COUR, L.F. LA, WELLS, B.: Fine structure and staining behaviour of heterochromatic segments in two plants. J. Cell Sci. **14**, 505–521 (1974).

COURT BROWN, W.M.: Human population cytogenetics. Amsterdam: North Holland Publ. 1967.

COURT BROWN, W.M., BUCKTON, K.E., JACOBS, P.A., TOUGH, J.H., V. KUENSSBERG, E., KNOX, J.D.E.: Chromosome studies in adults. Eugenics Lab. Mem. Ser. 42, Galton Laboratory. London: Cambridge Univ. Press 1966.

COURT BROWN, W.M., HARNDEN, D.G., JACOBS, P.A., MACLEAN, N., MANTLE, D.J.: Abnormalities of he sex chromosome complement in man. Med. Res. Counc. Spec. Rep. No. **305**, London: Her Majesty's Stationary Office 1964.

CRAIG-HOLMES, A.P., MOORE, F.B., SHAW, M.W.: Polymorphism of human C-band heterochromatin. I. Frequency of variants. Amer. J. hum. Genet. **25**, 181–192 (1973).

CRICK, F.: General model for the chromosomes of higher organisms. Nature (Lond.) **234**, 25–27 (1971).

CROSSEN, P.E.: Factors influencing Giemsa band formation of human chromosomes. Histochemie **35**, 51–62 (1973).

CZAKER, R.: Heterochromatin—ein möglicher Evolutionsmechanismus bei Amphibien. Verh. anat. Ges. 69. Vers. (1975, im Druck).

DARLINGTON, C.D.: The external mechanics of the chromosomes. Proc. roy. Soc. B **121**, 264–319 (1936).

DARLINGTON, C.D.: The chromosomes as a physico-chemical entity. Nature (Lond.) **176**, 1139–1144 (1955).

DARLINGTON, C.D., COUR, L.F. LA: Nucleic acid starvation of chromosomes in *Trilium*. J. Genet. **40**, 185–213 (1940).

DARLINGTON, C.D., UPCOTT, M.B.: The measurement of packing and contraction in chromosomes. Chromosoma (Berl.) **1**, 23–32 (1939).

DAVIDSON, W.M., SMITH, D.R.: A morphological sex difference in the polymorphnuclear neutrophil leukocytes. Brit. med. J. **1954 II**, 6–7.

DAVIES, H.G.: Electron microscope observations on the organization of heterochromatin in certain cells. J. Cell Sci. **3**, 129–150 (1968).

DEAVEN, L.L., STUBBLEFIELD, E.: Segregation of chromosomal DNA in chinese hamster fibroblasts in vitro. Exp. Cell Res. **55**, 132–135 (1969).

Denver Report 1960: A proposed standard system of nomenclature of human mitotic chromosomes. Lancet **1960 I**, 1063–1065.

DEV, V.G., WABURTON, D., MILLER, O.J., MILLER, D.A., ERLANGER, B.F., BEISER, S.M.: Consistent pattern of binding of antiadenosine antibodies to human metaphase chromosomes. Exp. Cell Res. **74**, 289–293 (1972).

DONAHUE, R.P., BIAS, W.B., RENWICK, J.H., McKUSICK, V.A.: Probable assignment of the Duffy blood group locus to chromosome 1 in man. Proc. nat. Acad. Sci. (Wash.) **61**, 949–955 (1968).

DRETS, M.E., SHAW, M.W.: Specific banding patterns of human chromosomes. Proc. nat. Acad. Sci. (Wash.) **68**, 2073–2077 (1971).

DUTRILLAUX, B.: Nouveau système de marquage chromosomique. Les bands T. Chromosoma (Berl.) **41**, 395–402 (1973).

DUTRILLAUX, B., GROUCHY, J. DE, FINAZ, C., LEJEUNE, J.: Mise en évidence de la structure fine des chromosomes humains par digestion enzymatique (pronase en particulier). C.R. Acad. Sci. (Paris), Ser. D, **273**, 587–588 (1971).

DUTRILLAUX, B., LAURENT, C., COUTURIER, J., LEJEUNE, J.: Coloration des chromosomes humains par l'acridine orange après traitment par le 5 bromodeoxyuridine. C.R. Acad. Sci. (Paris) **276**, 3179–3181 (1973).

DUTRILLAUX, B., LEJEUNE, J.: Sur une nouvelle technique d'analyse du caryotype humain. C.R. Acad. Sci. (Paris) **272**, 2638–2640 (1971).

EDWARDS, J.H., YUNCKEN, C., RUSHTON, D.I., RICHARDS, S., MITTWOCH, U.: Three cases of triploidy in man. Cytogenetics **6**, 81–104 (1967).

ELLIS, J.R., PENROSE, L.S.: An enlarged satellite and multiple malformations in the same pedigree. Ann. hum. Genet. **25**, 159–162 (1961).

ELLISON, J.R., BARR, H.J.: Quinacrine fluorescence of specific chromosome regions. Late replication and high AT content in *Samoaia leonensis*. Chromosoma (Berl.) **36**, 375–390 (1972).

EMMERICH, G.: Elektronenmikroskopische Untersuchungen interchromosomaler fädiger Strukturen an menschlichen Lymphocyten in der Metaphase. Hum. Genet. **19**, 227–234 (1973).

ENGMANN, F.R.: The morphology of human mitotic chromosomes. Ph. D. Thesis, University of Cambridge, cit. after E.H.R. Ford (1973).

EPHRUSSI, B., SORIEUL, S.: Nouvelles observations sur l'hybridation in vitro de cellules de souris. C.R. Acad. Sci. (Paris) **254**, 181–182 (1962).

EPSTEIN, C.J.: Mammalian Oocytes: X chromosome activity. Science **163**, 1078–1079 (1969).

EPSTEIN, C.J.: Expression of the mammalian X-chromosome before and after fertilization. Science **175**, 1467–1468 (1972).

EVANS, H.J., BUCKLAND, R.A., PARDUE, M.L.: Location of the genes coding for 18S and 28S ribosomal RNA in the human genome, Chromosoma (Berl.) **48**, 405–426 (1974).

EVANS, H.J., FORD, C.E., LYON, M.F., GRAY, J.: DNA replication and genetic expression in female mice with morphologically distinguishable X-chromosomes. Nature (Lond.) **206**, 900–903 (1965).

EVANS, H.J., GOSPEN, J.R., MITCHELL, A.R., BUCKLAND, R.A.: Location of human satellite DNA's on the Y chromosome. Nature (Lond.) **251**, 346–347 (1974).

EVANS, H.J., ROSS, A.: Spotted centromeres in human chromosomes. Nature (Lond.) **249**, 861–862 (1974).

FERGUSON-SMITH, M.A., FERGUSON-SMITH, M.E., ELLIS, P.M., DICKSON, M.: The sites and relative frequencies of secondary constrictions in human somatic chromosomes. Cytogenetics **1**, 325–343 (1962).

FERGUSON-SMITH, M.A., HANDMAKER, S.D.: The association of satellited chromosomes with specific chromosomal regions in cultured somatic cells. Ann. hum. Genet. **27**, 143–156 (1963).

FINAZ, C., GROUCHY, J. DE: Le caryotype humain après traitment par l'α-chymotrypsine. Ann. Génét. **14**, 309–311 (1971).

FITZGERALD, P.H.: Differential contraction of large and small chromosomes in cultured leucocytes of man. Cytogenetics **4**, 65–73 (1965).

FLAMM, W.G., WALKER, P.M.B., McCALLUM, M.: Some properties of single strands isolated from the DNA of the nuclear satellite of the mouse (*Mus musculus*). J. molec. Biol. **40**, 423–443 (1969).

FLEMMING, W.: Zellsubstanz, Kern- und Zellteilung. Leipzig: Vogel 1882.

FORD, C.E., HAMERTON, J.L.: The chromosomes of man. Nature (Lond.) **178**, 1020–1023 (1956).

FORD, E.H.R.: Human Chromosomes. London-New York: Academic Press 1973.

FORD, E.H.R., WOOLLAM, D.M.M.: The fine structure of the sex vesicle and sex chromosome association in spermatocytes of mouse, Golden Hamster and Field Vole. J. Anat. (Lond.) **100**, 487–499 (1966).

FRACCARO, M., KAIJSER, K., LINDSTEN, J.: Somatic chromosome complement in continuously cultured cells of two individuals with gonadal dysgenesis. Ann. hum. Genet. **24**, 45–61 (1960).

FRENSTER, J.H.: Nuclear polyanions as de-repressor of synthesis of RNA. Nature (Lond.) **206**, 680–683 (1965).

GAGNÉ, R., LABERGE, C.: Specific cytological recognition of the heterochromatic segment of number 9 chromosome in man. Exp. Cell Res. **73**, 239–242 (1972).

GAGNÉ, R., LABERGE, C., TANGUAY, R.: Interphase association of human "Y-body" with nucleolus. Johns Hopk. med. J. **130**, 254–258 (1972).

GAGNÉ, R., LABERGE, C., TANGUAY, R.: Aspect cytologique et localisation intranucléaire de l'heterochromatine constitutive des chromosomes C 9 chez l'homme. Chromosoma (Berl.) **41**, 159–166 (1973).

GAGNÉ, R., LUCIANI, J.M., DEVICTOR-VUILLET, M., STAHL, A.: C9 heterochromatin during the first meiotic prophase of human foetal oocyte. Exp. Cell Res. **85**, 111–116 (1974).

GALL, J.G.: Chromosome fibers from an interphase nucleus. Science **139**, 120–121 (1963).

GALL, J.G.: Chromosome fibers studied by a spreading technique. Chromosoma (Berl.) **20**, 221–233 (1966).

GALL, J.G., COHEN, E.H., POLAN, M.L.: Repetitive DNA sequences in *Drosophila*. Chromosoma (Berl.) **33**, 319–344 (1971).

GALPERIN, H.: Etude de la distribution générale des 46 chromosomes dans les cellules humaines en métaphase. Hum. Genet. **6**, 118–130 (1968).

GANNER, E., EVANS, H.J.: The relationship between patterns of DNA replication and of quinacrine fluorescence in the human chromosome complement. Chromosoma (Berl.) **35**, 326–341 (1971).

GARTLER, S.M., CHEN, S.H., FIALKOW, P.J., GIBLETT, E.R., SINGH, S.: X chromosome inactivation in cells from an individual heterozygous for two X-linked genes. Nature (Lond.) New Biol. **236**, 149–150 (1972).

GARTLER, S.M., LISKAY, R.M., CAMPELL, B.K., SPARKES, R., GANT, N.: Evidence for two functional X chromosomes in human oocytes. Cell Different. **1**, 215–218 (1972).

GARTLER, S.M., LISKAY, R.M., GANT, N.: Two functional X chromosomes in human fetal oocytes. Exp. Cell Res. **82**, 464–466 (1973).

GEBHART, E., BAUER, D.: Inter- und intra-chromosomale Verteilung von Chromatidtranslokationen nach Einwirkung von Trenimon auf menschliche Leukocyten in vitro. Chromosoma (Berl.) **32**, 152–161 (1970).

GEITLER, L.: Chromosomenbau. Berlin: Borntraeger, 1938.

GEITLER, L.: Das Heterochromatin der Geschlechtschromosomen bei Hemiptera. Chromosoma (Berl.) **1**, 197–230 (1939).

GEITLER, L.: Endomitose und endomitotische Polyploidisierung. Protoplasmalogie **VI C**. Wien: Springer 1953.

GERAEDTS, J.P.M., PEARSON, P.L.: Fluorescent chromosome polymorphisms: frequencies and segregation in a Dutch population. Clin. Genet. **6**, 247–257 (1974).

GERHARDT, H.: Kondensation der Chromosomen des Menschen in der Mitose. Hum. Genet. **10**, 158–167 (1970).

GERMAN, J.L.: Identification and characterization of human chromosomes by DNA replication sequence. Symp. Int. Cell Biol. **3**, 191–207 (1964a).

GERMAN, J.L.: The pattern of DNA synthesis in the chromosomes of human blood cells. J. Cell Biol. **20**, 37–55 (1964b).

GERMAN, J.L.: Cytological evidence for crossing-over in vitro in human lymphoid cells. Science **144**, 298–301 (1964c).

GERMAN, J.L., ARCHIBALD, R., BLOOM, P.: Chromosomal breakage in a rare and probably genetically determined syndrome of man. Science **148**, 506–507 (1965).

GIANNELLI, F.: Human chromosome DNA synthesis. In: Monographs in human genetics, vol. 5 (L. BECKMANN, M. HAUGE, eds.). Basel: Karger 1970.

GIBSON, D.A.: Somatic homologue association. Nature (Lond.) **227**, 164–165 (1970).

GILBERT, C.W., MULDAL, S., LAJTHA, L.G., ROWLEY, J.: Time sequence of human chromosome duplication. Nature (Lond.) **195**, 869–873 (1962).

GLÄSS, E.: Das Problem der Genomsonderung in den Mitosen unbehandelter Rattenlebern. Chromosoma (Berl.) **8**, 468–492 (1957).

GOLOMB, H.M., BAHR, G.F.: Electron microscopy of human interphase nuclei. Determination of total dry mass and DNA-packing ratio. Chromosoma (Berl.) **46**, 233–245 (1974).

GREILHUBER, J.: Differential staining of plant chromosomes after hydrochloric acid treatments (Hy bands). Österr. Bot. Z. **122**, 333–351 (1973).

GRIFFITH, J.D.: Chromatin structure: deduced from a minichromosome. Science **187**, 1202–1203 (1975).

GROPP, A., CITOLER, P., GEISLER, M.: Karyotypvariation und Heterochromatinmuster bei Igeln (*Erinaceus* und *Hemiechinus*). Chromosoma (Berl.) **27**, 288–307 (1969).

GROPP, A., NATARAJAN, A.T.: Karyotype and heterochromatin of the Algerian hedgehog. Cytogenetics **11**, 259–269 (1972).

GROPP, A., ODUNJO, F.: Beobachtungen zur morphologischen Konkordanz homologer Chromosomen somatischer Zellen. Exp. Cell Res. **30**, 577–582 (1963).

GRUMBACH, M.M., MORISHIMA, A., TAYLOR, J.H.: Human sex chromosome abnormalities in relation to DNA replication and heterochromatization. Proc. nat. Acad. Sci. (Wash.) **49**, 581–589 (1963).

GRUNDMANN, E., STEIN, P.: Untersuchungen über die Kernstruktur in normalen Geweben und im Carcinom. Beitr. path. Anat. **125**, 54–76 (1961).

GUTHERZ, S.: Zur Kenntnis der Heterochromosomen. Arch. mikr. Anat. **69**, 491–514 (1907).

HAMERTON, J.: Human Cytogenetics. London-New York: Academic Press 1971.

HANNAH, A.: Localization and function of heterochromatin in *Drosophila melanogaster*. Advanc. Genet. **4**, 87–125 (1951).

HARNDEN, D.S.: The chromosomes. In: Recent Advances in Human Genetics (L.S. PENROSE, ed.). London: Churchill 1961.

HARRIS, H.: Cell Fusion. Oxford: Clarendon Press 1970.

HARRIS, H., WATKINS, J.F.: Hybrid cells derived from mouse and man: Artificial heterokaryons of mammalian cells from different species. Nature (Lond.) **205**, 640–646 (1965).

HARRIS, H., WATKINS, J.F., FORD, C.E., SCHOEFL, G.I.: Artificial heterokaryons of animal cells from different species. J. Cell Sci. **1**, 1–30 (1966).

HAY, E.D., REVEL, J.P.: The fine structure of the DNP component of the nucleus. J. Cell Biol. **16**, 29–51 (1963).

HEIDENHAIN, M.: Plasma und Zelle. In: Handbuch der Anatomie, Bd. 8 (K. V. BARDELEBEN, Hrsg.), S. 1–1110. Jena: Fischer 1907.

HEIL, B.: Spätreplizierendes X-Chromosom und Sexchromatinhäufigkeit beim Menschen. Hum. Genet. **9**, 64–74 (1970).

HEITZ, E.: Das Heterochromatin der Moose. Jb. wissensch. Botan. **69**, 762–818 (1928).

HEITZ, E.: Heterochromatin, Chromozentren, Chromomeren. Ber. dtsch. botan. Gesellsch. **47**, 274–284 (1929).

HEITZ, E.: Die somatische Heteropyknose bei *Drosophila melanogaster* und ihre genetische Bedeutung. Z. Zellforsch. **20**, 237–287 (1933).

HEITZ, E.: Über α- und β-Heterochromatin sowie Konstanz und Bau der Chromomeren bei *Drosophila*. Biol. Zbl. **54**, 588–609 (1934).

HEITZ, E.: Chromosomenstruktur und Gene. Z. indukt. Abstam. Vererbungsl. **70**, 402–447 (1935).

HEITZ, E., BAUER, H.: Beweis für die Chromosomennatur der Kernschleifen in den Knäuelkernen von *Bibio hortulanus* L. (Cytologische Untersuchungen an Dipteren I). Z. Zellforsch. **17**, 67–82 (1933).

HENDERSON, A.S., WARBURTON, D., ATWOOD, K.C.: Location of ribosomal DNA in the human chromosome complement. Proc. nat. Acad. Sci. (Wash.) **69**, 3394–3398 (1972).

HENEEN, W.K.: In situ analysis of normal and abnormal patterns of the mitotic apparatus in cultured Rat-Kangaroo cells. Chromosoma (Berl.) **29**, 88–117 (1970).

HENEEN, W.K.: In situ analysis of cultured Potorous cells. II. Labeling in relation to chromosome orientation during the development of bipolar spindles. Hereditas (Lund) **67**, 251–258 (1971).

HENEEN, W.K., NICHOLS, W.W.: Nonrandom arrangement of metaphase chromosomes in cultured cells of the Indian deer, *Muntiacus muntjak*. Cytogenetics **11**, 153–164 (1972).

HENEEN, W.K., NICHOLS, W.W., LEVAN, A., NORBY, E.: Polykaryocytosis and mitosis in a human cell line after treatment with measles virus. Hereditas (Lund) **64**, 53–84 (1970).

HENKING, H.: Untersuchungen über die ersten Entwicklungsvorgänge in den Eiern der Insekten. II. Über Spermatogenese und deren Beziehung zur Eientwicklung bei *Pyrrhocoris apterus* L. Z. wissensch. Zoolog. **51**, 685–736 (1891).

HENNIG, W.: Giant chromosomes. In: The Cell Nucleus, vol. 2 (H. BUSCH, ed.), p. 333–369. London-New York: Academic Press 1974.

HENNIG, W., WALKER, P.M.B.: Variations in the DNA from the two rodent families (Cricetidae and Muridae). Nature (Lond.) **225**, 915–919 (1970).

HERREROS, B., GIANNELLI, F.: Spatial distribution of old and new chromatid sub-units and frequency of chromatid exchanges in induced human lymphocyte endoreduplications. Nature (Lond.) **216**, 286–288 (1967).

HEUMANN, H.G.: Electron microscope observations of the organisation of chromatin fibers in isolated nuclei of rat liver. Chromosoma (Berl.) **47**, 133–146 (1974).

HILWIG, I., GROPP, A.: Decondensation of constitutive heterochromatin in L cell chromosomes by a benzimidazole compound (33258 Hoechst). Exp. Cell Res. **81**, 474–477 (1973).

HIMES, M.: An analysis of heterochromatin in Maize root tips. J. Cell Biol. **35**, 175–181 (1967).

HOO, J.J., CRAMER, H.: On the position of chromosomes in prepared mitosis figures of human fibroblasts. Hum. Genet. **13**, 166–170 (1971).

HOPPE, P.C., WHITTEN, W.K.: Does X chromosome inactivation occur during mitosis of first cleavage? Nature (Lond.) **239**, 520 (1972).

HORSTMANN, E.: Zur Struktur des Nucleolus und des Kernes. Z. Zellforsch. **46**, 100–107 (1957).

HORSTMANN, E.: Die Kerneinschlüsse im Nebenhodenepithel des Hundes. Z. Zellforsch. **65**, 770–776 (1965).

HORSTMANN, E., RICHTER, R., ROOSEN-RUNGE, E.: Zur Elektronenmikroskopie der Kerneinschlüsse im menschlichen Nebenhodenepithel. Z. Zellforsch. **69**, 69–79 (1966).

HOWARD, A., PELC, S.R.: Synthesis of deoxyribonucleic acid in normal irradiated cells and its relation to chromosome breakage. Hereditas (Lund) Suppl. **6**, 261–273 (1953).

HSU, T.C.: Mammalian chromosomes in vitro. I. The karyotype of man. J. Hered. **43**, 167–172 (1952).

HSU, T.C.: Differential rate in RNA synthesis between euchromatin and heterochromatin. Exp. Cell Res. **27**, 332–334 (1962).

HSU, T.C.: Heterochromatin pattern in metaphase chromosomes of *Drosophila melanogaster*. J. Hered. **62**, 285–287 (1971).

HSU, T.C., ARRIGHI, F.: Distribution of constitutive heterochromatin in mammalian chromosomes. Chromosoma (Berl.) **34**, 243–253 (1971).

HSU, T.C., PATHAK, S., SHAFER, D.A.: Induction of chromosome crossbanding by treating cells with chemical agents before fixation. Exp. Cell Res. **79**, 484–487 (1973).

HSU, T.C., ZENZES, M.T.: Mammalian chromosomes in vitro. XVII. Idiogram of Chinese Hamster. J. nat. Cancer Inst. **32**, 857–869 (1964).

HUANG, R.C., BONNER, J.: Histone, a suppressor of chromosomal RNA synthesis. Proc. nat. Acad. Sci. (Wash.) **48**, 1216–1222 (1962).

HUGHES, A.: Some effects of abnormal tonicity on dividing cells in chick tissue cultures. Quart. J. micr. Sci. **93**, 207–219 (1952).

HUGHES-SCHRADER, S.: The chromosome cycle of *Planococcus* (Coccoidae). Biol. Bull. **69**, 462–468 (1935).

HUGHES-SCHRADER, S.: Cytology of coccids (Coccoidae, Homoptera). Advanc. Genet. **2**, 127–203 (1948).

HUNGERFORD, D.A.: Chromosome structure and function in man. I. Pachytene mapping in the male, improved methods and general discussion of initial results. Cytogenetics **10**, 23–32 (1971).

HUNGERFORD, D.A., ASHTON, F.T., BALABAN, G.B., LABADIE, G.U., MESSATZZIA, L.R., HALLER, G., MILLER, A.E.: The C-group pachytene bivalent with a locus characteristic for parachromosomally situated particulate bodies (parameres): A provisional map in human males. Proc. nat. Acad. Sci. (Wash.) **69**, 2165–2168 (1972).

HUNGERFORD, D.A., LABADIE, G.U., BALABAN, G.B.: Chromosome structure and function in man. II. Provisional maps of the two smallest autosomes (chromosomes 21 and 22) at pachytene in the male. Cytogenetics **10**, 33–37 (1971a).

HUNGERFORD, D.A., LaBADIE, G.U., BALABAN, G.B., MESSATZZIA, L.R., HALLER, G., MILLER, A.E.: Chromosome structure and function in man. IV. Provisional maps of the three long acrocentric autosomes (chromosomes 13, 14 and 15) at pachytene in the male. Ann. Génét. **14**, 157–260 (1971 b).

HUSKINS, C.L.: Nuclear Reproduction. In: Internat. Rev. Cytol., vol. 1 (G.H. BOURNE, J.F. DANIELLI, eds.), p. 9–26. London-New York: Academic Press 1952.

IKUSHIMA, T., WOLFF, S.: Sister chromatid exchanges induced by light flashes to 5-bromodeoxyuridine-and 5-iododeoxyuridine substituted Chinese Hamster chromosomes. Exp. Cell Res. **87**, 15–19 (1974).

IORIO, R.J., WYANDT, H.E.: Quinacrine studies of sex chromatin and nucleoli in human brain. Hum. Genet. **20**, 329–333 (1973).

JACOB, F., RYTER, A., CUZIN, F.: On the association between DNA and membrane in bacteria. Proc. roy. Soc. B **164**, 267–278 (1966).

JACOBJ, W.: Über das rhythmische Wachstum der Zellen durch Verdoppelung ihres Volumens. Arch. Entromech. Organ. **106**, 124–192 (1925).

JAFFRAY, J.Y., GENEIX, A.: Interchromosomal fibres: human ultrastructural study by a recent technique. Hum. Genet. **25**, 119–126 (1974).

JOHNSON, R.T., RAO, P.N.: Mammalian cell fusion: Induction of premature chromosome condensation in interphase nuclei. Nature (Lond.) **226**, 717–722 (1970).

JONES, K.W.: Chromosomal and nuclear location of mouse satellite DNA in individual cells. Nature (Lond.) **255**, 912–915 (1970).

JONES, K.W., CORNEO, G.: Location of satellite and homogeneous DNA sequences on human chromosomes. Nature (Lond.) New Biol. **233**, 268–271 (1971).

JONES, K.W., PROSSER, J., CORNEO, G., GINELLI, E., BOBROW, M.: Satellite DNA, constitutive heterochromatin, and human evolution. In: Modern Aspects of Cytogenetics: Constitutive heterochromatin in man. (R.A. PFEIFFER, ed.). Stuttgart-New York: Schattauer 1973 a.

JONES, K.W., PROSSER, J., CORNEO, G., GINELLI, E.: The chromosomal location of human satellite DNA III. Chromosoma (Berl.) **42**, 445–451 (1973 b).

JONES, K.W., PURDOM, I.F., PROSSER, J., CORNEO, G.: The chromosomal localization of human satellite DNA I. Chromosoma (Berl.) **49**, 161–171 (1974).

JUNKER, H.: Cytologische Untersuchungen an den Geschlechtsorganen der halbzwitterigen Steinfliege *Perla marginata*. Arch. Zellforsch. **17**, 185–259 (1923).

KATO, H., YOSIDA, T.H.: Banding pattern of Chinese hamster chromosomes revealed by new techniques. Chromosoma (Berl.) **36**, 272–280 (1972).

KAUFMANN, B.P., GAY, H., McDONALD, M.: Organizational patterns within chromosomes. In: Int. Rev. Cytol. vol. 9 (G.H. BOURNE, J.F. DANIELLE, eds.), p. 77–127. London-New York: Academic Press 1960.

KEYL, H.G., PELLING, C.: Differentielle DNS-Replikation in den Speicheldrüsen-Chromosomen von *Chironomus thummi*. Chromosoma (Berl.) **14**, 347–359 (1963).

KIM, M.A.: Identification and characterization of heterochromatic regions in the human metaphase and interphase nucleus. Hum. Genet. **21**, 331–340 (1974).

KIT, S.: Equilibrium sedimentation in density gradients of DNA preparations from animal tissues. J. molec. Biol. **3**, 711–716 (1961).

KLEINSCHMIDT, A.D., LANG, R., ZAHN, K.: Über Desoxyribonukleinsäure-Molekeln in Protein-Mischfilmen. Z. Naturforsch. **14 b**, 770 (1959).

KLINGER, H.P.: The fine structure of the sex chromatin body. Exp. Cell Res. **14**, 207–211 (1958).

KLINGER, H.P., DAVIS, J., GOLDHUBER, P., DITTA, T.: Factors influencing mammalian X chromosome condensation and sex chromatin formation. I. The effect of in vitro cell density on sex chromatin frequency. Cytogenetics **7**, 39–57 (1968).

KLINGER, H.P., LINDSTEN, J., FRACCARO, M., BARRAI, L., DOLINAR, Z.J.: DNA content and area of sex chromatin in subjects with structural and numerical aberrations of the X chromosome. Cytogenetics **4**, 96–116 (1965).

KLINGER, H.P., SCHWARZACHER, H.G.: The sex chromatin and heterochromatic bodies in human diploid and polyploid nuclei. J. biophys. biochem. Cytol. **8**, 345–364 (1960).

KLINGER, H.P., SCHWARZACHER, H.G., WEISS, J.: DNA content and size of sex chromatin positive female nuclei during the cell cycle. Cytogenetics **6**, 1–19 (1967).

KORENBERG, J.R., FREEDLENDER, E.F.: Giemsa technique for the detection of sister chromatid exchanges. Chromosoma (Berl.) **48**, 355–360 (1974).

KORNBERG, R.D.: Chromatin structure: A repeating unit of histones and DNA. Science **184**, 868–871 (1974).

KOSTANECKI, K. V.: Über parthenogetische Entwicklung der Eier von *Mactra* mit vorausgegangener oder unterbliebener Ausstoßung der Richtungskörper. Arch. mikr. Anat. **78**, 1–62 (1911).

KOULISCHER, L.: Le cyle de spiralisation des chromosomes mitotiques humains. Arch. Biol. (Liège) **74**, 391–413 (1963).

KOWARZYK, H., STEINHAUS, H., SZYMANIEC, S.: Arrangement of chromosomes in human cells. II. Distribution in metaphase figure. Bull. Acad. pol. Sci. **VI/14**, 401–404 (1966).

KUCHERLAPATI, R.S., CREAGAN, R.P., RUDDLE, F.H.: Progress in human gene mapping by somatic cell hybridization. In: The cell nucleus, vol. 2 (H. BUSCH, ed.). London-New York: Academic Press 1974.

LAIRD, C.D.: Chromatid structure: Relationship between DNA content and nucleotide sequence diversity. Chromosoma (Berl.) **32**, 378–406 (1971).

LAMPERT, F.: Feinstruktur und Trockengewicht menschlicher Chromosomen. Quantitative Elektronenmikroskopie. Naturwissenschaften **56**, 629–633 (1969).

LAMPERT, F., LAMPERT, P.: Ultrastructure of the human chromosome fiber. Hum. Genet. **11**, 9–17 (1970).

LATT, S.A.: Microfluorometric detection of deoxyribonucleic acid replication in human metaphase chromosomes. Proc. nat. Acad. Sci. (Wash.) **70**, 3395–3399 (1973).

LATT, S.A.: Microfluorometric analysis of DNA replication in human X chromosomes. Exp. Cell Res. **86**, 412–415 (1974).

LATT, S.A., DAVIDSON, R.L., LIN, M.S., GERALD, P.S.: Lateral asymmetry in the fluorescence of human Y-chromosomes stained with 33258 Hoechst. Exp. Cell Res. **87**, 425–429 (1974).

LAWLEY, P.D., CRATHORN, A.R., SHAH, S.A., SMITH, B.A.: Biomethylation of deoxyribonucleic acid in cultured human tumor cells (HeLa). Methylated bases other than 5-methylcytosin not detected. Biochem. J. **128**, 133–138 (1972).

LEE, C.S., THOMAS, C.A.: Formation of rings from *Drosophila* DNA fragments. J. molec. Biol. **77**, 25–55 (1973).

LEISTI, J.: Structural variation in human mitotic chromosomes. Ann. Acad. Sci. fenn. A, IV Biologica **179**, 1–69 (1971).

LENG, M., FELSENFELD, G.: The preferential interactions of polylysine and polyarginine with specific base sequences in DNA. Proc. nat. Acad. Sci. (Wash.) **56**, 1325–1332 (1966).

LEVAN, A.: Heterochromacy in chromosomes during their contraction phase. Hereditas (Lund) **32**, 449–468 (1946).

LEVAN, A., FREDGA, K., SANDBERG, A.: Nomenclature for centromeric position on chromosomes. Hereditas (Lund) **52**, 201–220 (1964).

LEVAN, A., HAUSCHKA, T.S.: Endomitotic reduplication mechanisms in ascites tumors of the mouse. J. nat. Cancer Inst. **14**, 1–40 (1953).

LEVAN, A., HSU, T.C., STICH, H.F.: The idiogram of the mouse. Hereditas (Lund) **48**, 677–687 (1962).

LEWIS, E.B.: The phenomenon of position effect. Advanc. Genet. **3**, 73–115 (1950).

LEWIS, W.H.: Observations on cells in tissue cultures with darkfield illumination. Anat. Rec. **26**, 15–29 (1923).

LICZNERSKI, G., LINDSTEN, J.: Trisomy 21 in man due to maternal non-disjunction during the first meiotic division. Hereditas (Lund) **70**, 153–154 (1972).

LIMA DE FARIA, A.: Differential uptake of tritiated thymidine into hetero—and euchromatin in *Melanoplus* and *Secale*. J. biophys. biochem. Cytol. **6**, 457–466 (1959).

LIMA DE FARIA, A., BIRNSTIEL, M., JAWORSKA, H.: Amplification of ribosomal cistrons in the heterochromatin of *Acheta*. Genetics, Suppl. **61**, 145–159 (1969).

LINDER, D., GARTLER, S.M.: Glucose-6-phosphate dehydrogenase mosaicism: utilization as a cell marker in the study of leiomyomas. Science **150**, 67–69 (1965).

LITTAU, V.C., BURDICK, C.J., ALLFREY, K.G., MIRSKY, A.E.: The role of histones in the maintenance of chromatin structure. Proc. nat. Acad. Sci. (Wash.) **54**, 1204–1212 (1965).

LITTLEFIELD, L.G., GOH, K.O.: Cytogenetic studies in control men and women. I. Variations in aberration frequencies in 29 709 metaphases from 305 cultures obtained over a three-year period. Cytogenet. Cell Genet. **12**, 17–34 (1973).

London Report 1963: The London Conference on the normal human karyotype. Cytogenetics **2**, 264–268 (1963).

154   Reference

Lubs, H.A., Ruddle, F.H.: Chromosome polymorphism in American negro and white populations. Nature (Lond.) **233**, 134–136 (1971).

Lyon, M.F.: Gene action in the X-chromosome of the mouse (*Mus musculus* L.). Nature (Lond.) **190**, 372 (1961).

Lyon, M.F.: Sex chromatin and gene action in the mammalian X-chromosome. Amer. J. hum. Genet. **14**, 135–148 (1962).

Lyon, M.F.: Chromosomal and subchromosomal inactivation. Ann. Rev. Genet. **2**, 31–52 (1968).

Lyon, M.F.: X-chromosome inactivation and developmental patterns in mammals. Biol. Rev. **47**, 1–35 (1972).

Macgregor, H.C., Walker, M.H.: The arrangement of chromosomes in nuclei of sperm from plethodontid Salamanders. Chromosoma (Berl.) **40**, 243–262 (1973).

Madan, K., Bobrow, M.: Structural variation in chromosome No. 9. Ann. Génét. **17**, 81–86 (1974).

Mandel, M., Marmur, J.: Use of ultraviolet absorbance-temperature profile for determining the guanine plus cytosine content of DNA. In: Methods in Enzymology, vol. 12 B (L. Grossmann, M. Moldave, eds.), p. 195–206. London-New York: Academic Press 1968.

Mars, R. De: Sex chromatin mass in living cultivated human cells. Science **138**, 980–981 (1962).

Martin, G.M., Sprague, C.A.: Parasexual cycle in cultivated human somatic cells. Science **166**, 761–763 (1969).

Matsui, S., Yoshida, H., Weinfeld, H., Sandberg, A.A.: Induction of prophase in interphase nuclei by fusion with metaphase cells. J. Cell Biol. **34**, 120–132 (1972).

Matthey, R.: Les chromosomes sexuels géants de *Microtus agrestis* L. Cellule **53**, 161–183 (1950).

Matthey, R.: Etudes sur les chromosomes *d'Ellobius lutescens* (Mammalia-Muridae-Microtinae). 1. Essai critique sur la valeur des critères proposés par le "Système Denver" pour l' identification des chromosomes homologues. Cytogenetics **1**, 180–195 (1962).

Maul, G.G., Price, J.W., Liebermann, M.W.: Formation and distribution of nuclear pore complexes in interphase. J. Cell Biol. **51**, 405–418 (1971).

Mayzel, W.: Über eigentümliche Vorgänge bei der Teilung der Kerne in Epithelialzellen. Zbl. med. Wiss. **13**, 849–852 (1875).

Mazia, D.: Mitosis and the Physiology of Cell Division. In: The Cell (J. Brachet, A.E. Mirsky, eds.). London-New York: Academic Press 1961.

McKay, R.D.G.: The mechanism of G and C banding in mammalian metaphase chromosomes. Chromosoma (Berl.) **44**, 1–14 (1973).

Meisner, L.F., Chuprevich, T.W., Inhorn, S.L.: Giemsa banding specifity. Nature (Lond.) New Biol. **245**, 145–147 (1973).

Melander, Y.: Chromosomal behavior during the origin of sex chromatin in the rabbit. Hereditas (Lund) **48**, 645–661 (1962).

Metz, C.W.: Chromosome studies on the Diptera. II. The paired association of chromosomes in the Diptera and its significance. J. exp. Zool. **21**, 213–279 (1916).

Mikelsaar, A.V.N., Tüür, S.J., Käosaar, M.E.: Human karyotype polymorphism. I. Routine and fluorescence microscopic investigation of chromosomes in a normal adult population. Hum. Genet. **20**, 89–101 (1973).

Mikelsaar, A.V.N., Wiikmaa, M.M., Tüür, S.J., Käosaar, M.E.: Human karyotype polymorphism. II. The distribution of individuals according to the presence of brilliant bands in chromosomes 3, 4 and 13 in a normal adult population. Hum. Genet. **23**, 59–63 (1974).

Miller, O.J.: Autoradiography in human cytogenetics. In: Advances in Human Genetics, vol. 1 (H. Harris, K. Hirschhorn, eds.), p. 35–130. London-New York: Plenum Press 1970.

Miller, O.J., Mukherjee, B.B., Breg, W.R., Gamble, A. van: Non-random distribution of chromosomes in metaphase figures from cultured human leucocytes. I. The peripheral location of the Y-chromosome. Cytogenetics **2**, 1–14 (1963a).

Miller, O.J., Breg, W.R., Mukherjee, B.B., Gamble, A. van: Non-random distribution of chromosomes in metaphase figures from cultured human leucocytes. II. The peripheral location of chromosomes 13, 17–18, and 21. Cytogenetics **2**, 151–167 (1963b).

Miller, O.J., Schnedl, W., Allen, J., Erlanger, B.F.: 5-Methylcytosine localised in mammalian constitutive heterochromatin. Nature (Lond.) **251**, 636–637 (1974).

Miller, O.L., Beatty, B.R.: Visualization of nucleolar genes. Science **164**, 955–957 (1969).

Mittwoch, U.: Barr bodies in relation to DNA values and nuclear size in cultured human cells. Cytogenetics **6**, 38–50 (1967).

Mittwoch, U.: Sex Chromosomes. London-New York: Academic Press 1967.

MONTGOMERY, T.H.: The terminology of aberrant chromosomes and their behavior in certain Hemiptera. Science 23, 36–38 (1906).

MOORE, K.L.: The Sex Chromatin. Philadelphia-London: Saunders 1966.

MOORHEAD, P.S., DEFENDI, V.: Asynchrony of DNA synthesis in chromosomes of human diploid cells. J. Cell Biol. 16, 202–209 (1963).

MOORHEAD, P.S., NOWELL, P.C., MELLMANN, W.J., BATTIPS, D.M., HUNGERFORD, D.A.: Chromosome preparations of leucocytes cultures from human peripheral blood. Exp. Cell Res. 20, 613–616 (1960).

MORISHIMA, A., GRUMBACH, M.M., TAYLOR, J.H.: Asynchronous duplication of human chromosomes and the origin of sex chromatin. Proc. nat. Acad. Sci. (Wash.) 48, 756–763 (1962).

MÜLLER, H.A.: Die Chromozentren in den Leberzellkernen der Maus unter normalen und pathologischen Bedingungen. Ergebn. allg. Pathol. pathol. Anat. 47, 145–185 (1966).

MÜLLER, H.J., KLINGER, H.P.: Chromosome polymorphism in a human newborn population. In: Chromosomes Today 5 (in press) (1975).

MÜLLER, W., ROSENKRANZ, W.: Rapid banding technique for human and mammalian chromosomes. Lancet 1972 I, 898.

MULLER, H.J.: Variegation in Drosophila and the inert chromosome regions. Proc. nat. Acad. Sci. (Wash.) 22, 27–33 (1936).

MULLER, H.J., PAINTER, T.S.: The differentiation of the sex chromosomes of Drosophila into genetically active and inert regions. Z. indukt. Abstamm.- u. Vererb.-L. 62, 316–365 (1932).

NAGL, W.: Correlation of chromatin structure and interphase stage in nuclei of Allium flavum. Cytobiologie 1, 395–398 (1970).

NARDI, I., RAGGHIANTI, M., MANCINO, G.: Banding patterns in newt chromosomes by the Giemsa stain. Chromosoma (Berl.) 40, 321–331 (1973).

NATARAJAN, A.T., GROPP, A.: A fluorescence study of heterochromatin and nucleolar organization in the laboratory and tobacco mouse. Exp. Cell Res. 74, 245–250 (1972).

NATARAJAN, A.T., SHARMA, R.P., AHNSTRÖM, G.: Fluorochromes and heterochromatin. A study of the chromosomes of Microtus agrestis L. Hereditas (Lund) 69, 217–222 (1971).

NEBEL, B.R.: On the structure of mammalian chromosomes during spermatogenesis and after radiation, with special reference to cores. Proc. IVth Int. Conf. Electron Microscop. Vol. 2, p. 227. Berlin-Göttingen-Heidelberg: Springer 1958.

New Haven Conference (1973): First International Workshop on Human Gene Mapping. Birth Defects: Original Article Series X:3. The National Foundation, New York 1974.

NOWELL, P.C.: Mitotic inhibition and chromosome damage by mitomycin C on human leukocytes. Exp. Cell Res. 33, 445–449 (1964).

NUR, U.: Nonreplication of heterochromatic chromosomes in a mealy bug. Planococcus citri (Coccoidea: Homoptera). Chromosoma (Berl.) 19, 439–448 (1966).

O'BRIEN, R.L., SANYAL, A.B., STANTON, R.H.: DNA replication sites in HeLa cells. Exp. Cell Res. 80, 340–344 (1973).

OCKEY, C.H.: The position of chromosomes at metaphase in human fibroblasts and their DNA synthesis behaviour. Chromosoma (Berl.) 27, 308–320 (1969).

OHNO, S.: Cytologic and genetic evidence of somatic segregation in mammals, birds and fishes. In: Phenotypic expression in vitro. Vol. 2. Baltimore: Tissue Culture Ass., Williams and Wilkins 1966.

OHNO, S.: Sex chromosomes and sex-linked genes. Berlin-Heidelberg-New York: Springer 1967.

OHNO, S.: So much "junk" DNA in our genome. In: Evolution of Genetic Systems, vol. 23 (H.H. SMITH, ed.), Brookhaven Symposia, p. 366–370. New York: Gordon and Breach 1972.

OHNO, S.: Protochordata, Cyclostomata and Pisces. In: Animal Cytogenetics, vol. 4 (B. JOHN, H. BAUER, S. BROWN, H. LAYANA, A. LEVAN, M. WHITE, eds.), Chordata 1. Berlin-Stuttgart: Borntraeger 1974.

OHNO, S., CATTANACH, B.M.: Cytological study of X-autosome translocation in Mus musculus. Cytogenetics 1, 129–190 (1962).

OHNO, S., KAPLAN, W.D., KINOSITA, R.: Formation of the sex chromatin by a single X-chromosome in liver cells of Rattus norvegicus. Exp. Cell Res. 18, 415–418 (1959).

OHNO, S., KAPLAN, W.D., KINOSITA, R.: X-chromosome behaviour in germ and somatic cells of Rattus norvegicus. Exp. Cell Res. 22, 535–544 (1961).

OHNO, S., KLINGER, H.P., ATKINS, N.B.: Human oogenesis. Cytogenetics 1, 42–51 (1962).

OHNO, S., MAKINO, S.: The single-X nature of sex-chromatin in man. Lancet 1961 I, 78–79.

OHNO, S., TRUJILLO, J.M., KAPLAN, J.M., KINOSITA, R.: Nucleolus organizers in the causation of chromosomal anomalies in man. Lancet **1961 II**, 123.

OHNUKI, Y.: Demonstration of the spiral structure of human chromosomes. Nature (Lond.) **203**, 916–917 (1965).

OHNUKI, Y.: Structure of chromosomes I. Morphological studies of the spiral structure of human somatic chromosomes. Chromosoma (Berl.) **25**, 402–428 (1968).

OLINS, A.L., OLINS, D.E.: Spheroid chromatin units (ν-bodies). Science **183**, 330–332 (1974).

OSGOOD, E.E., JENKINS, D.P., BROOKS, R., LAWSON, R.K.: Electron micrographic studies of the expanded and uncoiled chromosomes from human leucocytes. Ann. N.Y. Acad. Sci. **113**, 717–726 (1964).

ÖSTERGREN, G.: Isopycnosis and isopycnotic; two new terms for use in chromosome studies. Hereditas (Lund) **36**, 511–513 (1950).

PACHMANN, U., RIGLER, R.: Quantum yield of acridines interacting with DNA of defined base sequence. A basis for the explanation of acridine bands in chromosomes. Exp. Cell Res. **72**, 602–608 (1972).

PALITTI, F., RIZZONI, M.: Experimental evolution of cell populations of Chinese hamster treated with colchicine. Induction of high degree of ploidy and a phasespecific lethal effect. Int. J. Cancer **9**, 510–523 (1972).

PARDON, J.F., WILKINS, M.H.F.: A super-coil model for nucleohistone. J. molec. Biol. **68**, 115–124 (1972).

PARDON, J.F., WILKINS, M.H.F., RICHARDS, B.M.: Molecular structure super helical model for nucleohistone. Nature (Lond.) **215**, 508–509 (1967).

PARDUE, M.L., GALL, J.G.: Molecular hybridisation of radioactive DNA to the DNA of cytological preparations. Proc. nat. Acad. Sci. (Wash.) **64**, 600–604 (1969).

PARDUE, M.L., GALL, J.G.: Chromosomal localization of mouse satellite DNA. Science **168**, 1356–1358 (1970).

PARDUE, M.L., GALL, J.G.: Molecular cytogenetics. In: Molecular Genetics and Developmental Biology (M. SUSSMANN, ed.). Englewood Cliffs: Prentice Hall 1972.

Paris Conference (1971): Standardization in Human Cytogenetics. Birth Defects: Original Article Series **VIII, 7**, The National Foundation, New York 1972.

PARK, W.W.: The occurrence of sex chromatin in early human and macaque embryos. J. Anat. (Lond.) **91**, 369–373 (1957).

PASSARGE, E.: The human karyotype. Analysis of Chromosomes in Mitosis and Evaluation of Cytogenetic Data. In: Methods in Human Cytogenetics (H.G. SCHWARZACHER, U. WOLF, eds.), Berlin-Heidelberg-New York: Springer 1974.

PASSARGE, E., FRIES, E.: X-chromosome inactivation in X-linked hypohidrotic ectodermal dysplasia. Nature New Biol. **245**, 58–59 (1973).

PATAU, K.: Identification of chromosomes. In: Human Chromosome Methodology (J.J. YUNIS, ed.). London-New York: Academic Press 1965.

PAWELETZ, N.: Elektronenmikroskopische Untersuchungen an frühen Stadien der Mitose bei HeLa-Zellen. Cytobiologie **9**, 368–390 (1974).

PAWLOWITZKI, I.H., BLASCHKE, R., CHRISTENHUSS, R.: Darstellung von Chromosomen im Raster-Elektronenmikroskop nach Enzymbehandlung. Naturwissenschaften **55**, 63–64 (1968).

PAWLOWITZKI, I.H., CENANI, A.: Sporadic triploid cells in human blood and fibroblast cultures. Hum. Genet. **5**, 65–69 (1967).

PAWLOWITZKI, I.H., PEARSON, P.L.: Chromosomal aneuploidy in human spermatozoa. Hum. Genet. **16**, 119–122 (1972).

PEARSON, L., BOBROW, M.: Definitive evidence for the short arm of the Y chromosome associating with the X chromosome during meiosis in the human male. Nature (Lond.) **226**, 959–961 (1970).

PEARSON, P.L., BOBROW, M., VOSA, C.G.: Technique for identifying Y chromosomes in human interphase nuclei. Nature (Lond.) **226**, 78–80 (1970).

PERA, F.: Dauer der DNS-Replikation von Eu- und Heterochromatin bei *Microtus agrestis*. Chromosoma (Berl.) **25**, 21–29 (1968).

PERA, F.: Struktur und Position der heterochromatischen Chromosomen in Interphasekernen von *Microtus agrestis*. Z. Zellforsch. **98**, 421–436 (1969).

PERA, F.: Mechanismen der Polyploidisierung und der somatischen Reduktion. Ergebn. Anat. Entwickl.-Gesch. 43/5, 1–112 (1970).

PERA, F., RAINER, B.: Studies of multipolar mitoses in euploid tissue cultures. I. Somatic reduction to exactly haploid and triploid chromosome sets. Chromosoma (Berl.) **42**, 71–86 (1973).

PERA, F., RAINER, B.: Studies of multipolar mitoses in euploid tissue cultures. II. Somatic segregation of the sex chromosomes. Chromosoma (Berl.) **46**, 225–232 (1974).

PERA, F., SCHOLZ, P.: Polyploidization in vitro: Formation of a predominantly triploid cell population in an originally diploid tissue culture of *Microtus agrestis*. Hum. Genet. **21**, 17–26 (1974).

PERA, F., SCHWARZACHER, H.G.: Die Verteilung der Chromosomen auf die Tochterzellkerne multipolarer Mitosen in euploiden Gewebekulturen von *Microtus agrestis*. Chromosoma (Berl.) **26**, 337–354 (1969).

PERA, F., SCHWARZACHER, H.G.: Lokalisation der heterochromatischen Chromosomen von *Microtus agrestis* in Interphase und Mitose. Cytobiologie **2**, 188–199 (1970).

PERA, F., WOLF, U.: DNS-Replikation und Morphologie der X-Chromosomen während der Syntheseperiode bei *Microtus agrestis*. Chromosoma (Berl.) **22**, 378–389 (1967).

PERRY, P., WOLFF, S.: New Giemsa method for the differential staining of sister chromatids. Nature (Lond.) **251**, 156–158 (1974).

PISCHINGER, A.: Untersuchungen über die Kernstruktur, besonders über die Beziehung zwischen Struktur im Leben und nach Fixierung. Z. Zellforsch. **26**, 249–280 (1937).

PISCHINGER, A.: Über die Struktur des Zellkernes. Protoplasma (Wien) **39**, 567–587 (1950).

PRAW, E.J. DU: Macromolecular organization of nuclei and chromosomes: A folded fibre model based on whole mount electron microscopy. Nature (Lond.) **206**, 338–343 (1965a).

PRAW, E.J. DU: The organization of nuclei and chromosomes in honeybee embryonic cells. Proc. nat. Acad. Sci. (Wash.) **53**, 161–169 (1965b).

PRAW, E.J. DU: Evidence for a "folded fibre" organization in human chromosomes. Nature (Lond.) **209**, 577–581 (1966).

PRAW, E.J. DU, BAHR, G.F.: The arrangement of DNA in human chromosomes, as investigated by quantitative electron microscopy. Acta cytol. (Philad.) **13**, 188–205 (1969).

PRESCOTT, D.M., BENDER, M.A.: Autoradiographic study of chromatid distribution of labeled DNA in two types of mammalian cells in vitro. Exp. Cell Res. **29**, 430–442 (1963).

PRIEST, J.H., HEADY, J.E., PRIEST, R.: Delayed onset of replication of human X-chromosomes. J. Cell Biol. **35**, 483–487 (1967).

PYERITZ, R.E., THOMAS, C.A.: Regional organization of eukaryotic DNA sequences as studied by the formation of folded rings. J. molec. Biol. **77**, 57–73 (1973).

RAO, P.N., JOHNSON, R.T.: Mammalian cell fusion: Studies on the regulation of DNA synthesis and mitosis. Nature (Lond.) **225**, 159–164 (1970).

RAO, P.N., JOHNSON, R.T.: Induction of chromosome condensation in interphase cells. In: Advances in Cell and Molecular Biology, vol. 3 (E.J. DU PRAW, ed.). London-New York: Academic Press 1974.

RAO, R.N., NATARAJAN, A.T.: Somatic association in relation to chemically induced chromosome aberrations in *Vicia faba*. Genetics **57**, 821–835 (1967).

RHOADES, M.M.: Studies of telocentric chromosome in maize with reference to the stability of its centromere. Genetics **25**, 483–520 (1940).

RICHARDS, B.M., PARDON, J.F.: The molecular structure of nucleohistone (DNH). Exp. Cell Res. **62**, 184–196 (1970).

RIGLER, R.: Microfluorometric characterization of intracellular nucleic acids and nucleoproteins by acridine orange. Acta physiol. scand. **67**, Suppl. 267, 1–122 (1966).

RIS, H.: Ultrastructure and molecular organization of genetic systems. Canad. J. Genet. Cytol. **3**, 95–120 (1961).

RIS, H.: Fine structure of chromosomes. Proc. roy Soc. B**164**, 246–254 (1966).

RIS, H.: The molecular organization of chromosomes. In: Handbook of Molecular Cytology (A. LIMA-DE-FARIA, ed.), 221–250. Amsterdam: North Holland Publ. 1969.

RIZZONI, M., PALITTI, F., PERTICONE, P.: Euploid segregation through multipolar mitosis in mammalian cell culture. Chromosoma (Berl.) **45**, 151–162 (1974).

ROBBINS, E., GONATAS, N.K.: The ultrastructure of a mammalian cell during the mitotic cycle. J. Cell Biol. **21**, 429–463 (1964).

ROBINSON, J.A.: Origin of extra chromosome in trisomy 21. Lancet **1973 I**, 131–133.

RODMAN, T.C., TAHILIANI, S.: The Feulgen banded karyotype of the mouse: Analysis of the mechanisms of banding. Chromosoma (Berl.) **42**, 37–56 (1973).

Röhme, D.: Prematurely condensed chromosomes of the Indian muntjac: A model system for the analysis of chromosome condensation and banding. Hereditas (Lund) 76, 251–258 (1974).

Rosenberg, O.: Über die Individualität der Chromosomen im Pflanzenreich. Flora 93, 251–259 (1904).

Rosenberg, O.: Über den Bau des Ruhekerns. Svensk. botan. Tid. 3, 163–173 (1909).

Rosenkranz, W., Fleck, S.: Die Bedeutung der Assoziation satellitentragender Chromosomen. Hum. Genet. 7, 9–21 (1969).

Ross, A., Gormley, I.P.: Examination of surface topography of Giemsa-banded human chromosomes by light and electron microscopic techniques. Exp. Cell Res. 81, 79–86 (1973).

Ruddle, F.H.: Utilization of somatic cells for genetic analysis: Possibilities and problems. Symp. Int. Soc. Cell Biol. 9, 233–264 (1970).

Ruddle, F.H.: Linkage analysis using somatic cell hybrids. In: Advances in Human Genetics, vol. 3 (H. Harris, K. Hirschhorn, eds.), p. 173–236. New York-London: Press 1972.

Rutishauser, A., Haemmerli, G., Sträuli, P.: Cytogenetik transplantabler tierischer und menschlicher Tumoren. Neujahrsblatt naturforsch. Ges. (Zürich) 107, 1–85 (1963).

Ruzicka, F.: Über die Primärwindungen menschlicher Chromosomen. Hum. Genet. 20, 335–341 (1973).

Ruzicka, F.: Effect of G-banding techniques on the ultrastructure of human chromosomes. Hum. Genet. 22, 119–126 (1974).

Ruzicka, F., Schwarzacher, H.G.: The ultrastructure of human mitotic chromosomes and interphase nuclei treated by Giemsa banding techniques. Chromosoma (Berl.) 46, 443–454 (1974).

Salomon, R., Kaye, A.M., Herzberg, M.: Mouse nuclear satellite DNA 5-methylcytosine content, pyrimidine isopyrolite distribution and electron microscopic appearence. J. molec. Biol. 43, 581–592 (1969).

Sanchez, O., Yunis, J.J.: The relationship between repetitive DNA and chromosomal bands in man. Chromosoma (Berl.) 48, 191–202 (1974).

Sandberg, A.A., Aya, T., Ikeuchi, T., Weinfeld, H.: Definition and morphologic features of chromosome pulverization: A hypothesis to explain the phenomenon. J. nat. Cancer Inst. 45, 615–623 (1970).

Sandritter, W., Kiefer, G., Schlüter, G., Moore, W.: Eine cytophotometrische Methode zur Objektivierung der Morphologie von Zellkernen. Ein Beitrag zum Problem von Eu- und Heterochromatin. Histochemie 10, 341–352 (1967).

Sasaki, M.: Observation on the modification in size and shape of chromosomes due to technical procedure. Chromosoma (Berl.) 11, 514–522 (1961).

Saunders, G.F., Hsu, T.C., Getz, M.J., Simes, E.L., Arrighi, F.E.: Location of a human satellite DNA in human chromosomes. Nature (Lond.) New Biol. 236, 244–246 (1972a).

Saunders, G.F., Shirakawa, S., Saunders, P., Arrighi, F.E., Hsu, T.C.: Populations of repeated DNA sequences in the human genome. J. molec. Biol. 63, 323–334 (1972b).

Sheid, B., Srinivasan, P.R., Borek, E.: Deoxyribonucleic acid methylase of mammalian tissues. Biochemistry 7, 280–285 (1968).

Schmid, W.: DNA replication patterns of human chromosomes. Cytogenetics 2, 175–193 (1963).

Schmid, W.: Multipolar spindles after endoreduplication. Exp. Cell Res. 42, 201–204 (1966).

Schmid, W.: Heterochromatin in mammals. Arch. Klaus-Stift. Vererb.-Forsch. 42, 1–60 (1967).

Schmid, W., Leppert, M.F.: Karyotyp, Heterochromatin und DNS-Werte bei 13 Arten von Wühlmäusen (Microtinae-Rodentia-Mammalia) Arch. Klaus-Stift. Vererb.-Forsch. 43, 88–91 (1968).

Schmid, W., Leppert, M.F.: Rates of DNA synthesis in heterochromatic and euchromatic segments of the chromosome complements of two rodents. Cytogenetics 8, 125–135 (1969).

Schmid, W., Smith, D.W., Theiler, K.: Chromatinmuster in verschiedenen Zelltypen und Lokalisation von Heterochromatin in Metaphasechromosomen bei Microtus agrestis, Mesocricetus auratus, Cavia cobaya und beim Menschen. Arch. Klaus-Stift. Vererb.-Forsch. 40, 35–49 (1965).

Schnedl, W.: Banding pattern of human chromosomes. Nature (Lond.) New Biol. 233, 93–94 (1971a).

Schnedl, W.: Analysis of the human karyotype using a reassociation technique. Chromosoma (Berl.) 34, 448–454 (1971b).

Schnedl, W.: Unterschiedliche Fluorescenz der beiden homologen Chromosomen Nr. 3 beim Menschen. Hum. Genet. 12, 59–63 (1971c).

SCHNEDL, W.: Fluorescenzuntersuchungen über die Längenvariabilität des Y-Chromosoms beim Menschen. Hum. Genet. **12**, 188–194 (1971 d).

SCHNEDL, W.: Giemsa banding, quinacrine fluorescence and DNA replication in chromosomes of cattle (*Bos taurus*). Chromosoma (Berl.) **38**, 319–328 (1972).

SCHNEDL, W.: Late DNA replication pattern of human chromosomes determined by means of ^{3}H-deoxycytidine. Hum. Genet. **20**, 55–68 (1973).

SCHNEDL, W.: Banding patterns in human chromosomes visualized by Giemsa staining after various pretreatments. In: Methods in Human Cytogenetics (H.G. SCHWARZACHER, U. WOLF, eds.). Berlin-Heidelberg-New York: Springer 1974 a.

SCHNEDL, W.: Der Polymorphismus des menschlichen Chromosomensatzes — eine Möglichkeit für den Vaterschaftsnachweis. Z. Rechtsmedizin **74**, 17–23 (1974 b).

SCHNEDL, W., CZAKER, R.: Centromeric heterochromatin and comparison of G-banding in cattle, goat, and sheep chromosomes (Bovidae). Cytogenet. Cell Genet. **13**, 246–255 (1974).

SCHNEIDER, E., HEUKAMP, U., PERA, F.: Loss of heteropycnosis of the constitutive heterochromatin in specifically activated cells of the thyroid gland of *Microtus agrestis*. Chromosoma (Berl.) **41**, 167–173 (1973).

SCHNEIDERMANN, L.J., SMITH, C.A.B.: Non-random distribution of certain homologous pairs of normal human chromosomes in metaphase. Nature (Lond.) **195**, 1229–1230 (1962).

SCHRADER, F.: Experimental and cytological investigations of the life cycle of *Gossiparia spuria* (Coccoidae) and their bearing on the problem of haploidy in males. Z. wiss. Zool. **134**, 149–179 (1929).

SCHRECK, R.R., WARBURTON, D., MILLER, O.J., BEISER, S.M., ERLANGER, B.F.: Chromosome structure as revealed by a combined chemical and immunochemical procedure. Proc. nat. Acad. Sci. (Wash.) **70**, 804–807 (1973).

SCHULTZ, J.: Variegation in *Drosophila* and the inert chromosome regions. Proc. nat. Acad. Sci. (Wash.) **22**, 27–33 (1936).

SCHULTZ, J.: The function of heterochromatin. Proc. 7th Intern. Congr. Genet., Edinburgh 1939. J. Genet. Suppl. 257–262 (1941).

SCHWARZACHER, H.G.: Sex chromatin in living human cells. Cytogenetics **2**, 117–128 (1963).

SCHWARZACHER, H.G.: Kernstruktur und Geschlechtschromatin menschlicher Zellen in vitro im lebenden Zustand und nach Fixierung. Acta anat. (Basel) **57**, 91–104 (1964).

SCHWARZACHER, H.G.: Sexchromatin in polyploiden Zellen. Hum. Genet. **2**, 28–35 (1966).

SCHWARZACHER, H.G.: Chromosome mosaics as markers in embryology. In: Comparative Mammalian Cytogenetics (K. BENIRSCHKE, ed.). Berlin-Heidelberg-New York: Springer 1969.

SCHWARZACHER, H.G.: Die Ergebnisse elektronenmikroskopischer Untersuchungen an somatischen Chromosomen des Menschen. Hum. Genet. **10**, 195–208 (1970).

SCHWARZACHER, H.G.: Analysis of interphase nuclei. In: Methods in Human Cytogenetics (H.G. SCHWARZACHER, U. WOLF, eds.). Berlin-Heidelberg-New York: Springer 1974.

SCHWARZACHER, H.G., RUZICKA, F., SPERLING, K.: Electron microscopy of human G-band stained metaphase and premature condensed chromosomes. In: Chromosomes today **5**, in press (1975).

SCHWARZACHER, H.G., SCHNEDL, W.: Der Zellzyklus in Fibroblastenkulturen vom Menschen. Z. Zellforsch. **67**, 165–173 (1965 a).

SCHWARZACHER, H.G., SCHNEDL, W.: Endoreduplication in human fibroblast cultures. Cytogenetics **4**, 1–18 (1965 b).

SCHWARZACHER, H.G., SCHNEDL, W.: Position of labelled chromatids in diplochromosomes of endoreduplicated cells after uptake of tritiated thymidine. Nature (Lond.) **209**, 107–108 (1966).

SCHWARZACHER, H.G., SCHNEDL, W.: Elektronenmikroskopische Untersuchungen menschlicher Metaphasen-Chromosomen. Hum. Genet. **4**, 153–165 (1967).

SCHWARZACHER, H.G., SCHNEDL, W.: Zur Ultrastruktur der Chromosomen des Menschen. Hum. Genet. **8**, 75–80 (1969).

SCHWARZACHER, H.G., WOLF, U. (Eds.): Methods in Human Cytogenetics. Berlin-Heidelberg-New York: Springer 1974.

SCHWEIZER, D.: Differential staining of plant chromosomes with Giemsa. Chromosoma (Berl.) **40**, 307–320 (1973).

SEABRIGHT, M.: The use of proteolytic enzymes for the mapping of structural rearrangements in the chromosomes of man. Chromosoma (Berl.) **36**, 204–210 (1972).

SEHESTED, J.: Giemsa "banding" in metaphase chromosomes after pretreatment with inactivated trypsin. Hum. Genet. **19**, 321–324 (1973).

SHAPIRO, H.S., CHARGAFF, E.: Studies on the nucleotide arrangement in deoxyribonucleic acids. IV. Patterns of nucleoside sequence in the deoxyribonucleic acid of rye germ and its fractions. Biochim. biophys. Acta (Amst.) **39**, 68–82 (1960).

SHAW, M.W., SCHWAB, L.E., BRINKLEY, B.R.: Electron microscopy of human chromosomes. In: Perspectives in Cytogenetics. The Next Decade (S.W. WRIGHT, B.F. CRANDALL, L. BOYER, eds.), p. 251–284. Springfield/Ill.: Thomas 1972.

SHIRAISHI, Y., YOSIDA, T.H.: Banding pattern analysis of human chromosomes by use of a urea treatment technique. Chromosoma (Berl.) **37**, 75–83 (1972).

SIEGER, M., PERA, F., SCHWARZACHER, H.G.: Genetic inactivity of heterochromatin and heteropycnosis in *Microtus agrestis*. Chromosoma (Berl.) **29**, 349–364 (1970).

SINHA, A.K.: Spontaneous occurrence of tetraploidy and nearhaploidy in mammalian peripheral blood. Exp. Cell Res. **47**, 443–448 (1967).

SLIZYNSKI, B.M.: Ectopic pairing and the distribution of heterochromatin in the X-chromosome of salivary gland nuclei of *Drosophila melanogaster*. Proc. roy Soc. (Edinb.) **62**, 114–122 (1945).

SMART, J.E., BONNER, J.: Studies on the role of histones in the structure of chromatin. J. molec. Biol. **58**, 661–674 (1971).

SMITH, S.G.: Heterochromatin, colchicine and karyotype. Chromosoma (Berl.) **16**, 162–165 (1965).

SOLARI, A.J.: The ultrastructure of chromatin fibers. Exp. Cell Res. **53**, 567–581 (1968).

SORSA, V.: Whole mount electron microscopy of core fibrils in salivary-gland chromosomes of *Drosophila melanogaster*. Hereditas (Lund) **72**, 169–172 (1972).

SORSA, V.: Condensation of chromosomes during mitotic prophase. Whole mount electron microscopy of cerebral ganglion cells of *Drosophila*. Hereditas (Lund) **75**, 101–108 (1973a).

SORSA, V.: Chromatin fibril size in salivary gland chromosomes of *Drosophila melanogaster*. Hereditas (Lund) **74**, 133–137 (1973b).

SORSA, V., SORSA, M.: Ideas on the lateral organization of chromosomes revived by an observation of four stranded mitotic prophase chromosome in *Hyacinthus*. Ann. Acad. Sci. fenn. A **133**, 1–11 (1968).

SORSA, V., SORSA, M., VIRRANKOSKI, V., PUSA, K.: An electron microscopy study on alkali-urea treated salivary gland chromosomes of *Drosophila*. Ann. Acad. Sci. fenn. **A IV** Biologica **166**, 1–10 (1970).

SOUTHERN, E.M.: Base sequence and evolution of guinea pig α-satellite DNA. Nature (Lond.) **277**, 794–798 (1970).

SPAETER, M.: Nicht zufällige Verteilung homologer Chromosomen (Nr. 9 und YY) in Interphasekernen menschlicher Fibroblasten. Hum. Genet. **27**, 111–118 (1975).

SPARVOLI, E., GAY, H., KAUFMANN, B.P.: Number and pattern of association of chromonemata in the chromosomes of *Tradescantia*. Chromosoma (Berl.) **16**, 415–435 (1965).

SPERLING, K., RAO, P.N.: Mammalian cell fusion: V. Replication behaviour of heterochromatin as observed by premature chromosome condensation. Chromosoma (Berl.) **45**, 121–131 (1974a).

SPERLING, K., RAO, P.N.: The phenomenon of premature chromosome condensation: Its relevance to basic and applied research. Hum. Genet. **23**, 235–258 (1974b).

SPERLING, K., WIESNER, R.: A rapid banding technique for routine use in human and comparative cytogenetics. Hum. Genet. **15**, 349–353 (1972).

STAHL, A., LUCIANI, J.M., GAGNÈ, R., DEICTOR, M., CAPODANO, A.M.: Heterochromatin, micronucleoli and RNA containing body in the diplotene stage of the human oocyte. Chromosomes today **5**, in press (1975).

STAMBROOK, P.J., FLICKINGER, R.A.: Changes in chromosomal DNA replication patterns in developing frog embryos. J. exp. Zool. **174**, 101–114 (1970).

STERN, C.: Ein genetischer und zytologischer Beweis für Vererbung im Y-Chromosom von *Drosophila melanogaster*. Z. indukt. Abstamm.- u. Vererb.-L. **44**, 187–231 (1927).

STERN, C.: Somatic crossing-over and segregation in *Drosophila melanogaster*. Genetics **21**, 625–730 (1936).

STERN, C.: The nucleus and somatic cell variation. J. cell comp. Physiol. **52**, Suppl. 1, 1–34 (1958).

STEVENS, L.J.: Electron microscopy of unsectioned human chromosomes. Ann. hum. Genet. **31**, 267–275 (1967).

STEVENS, N.M.: A study of the germ cells of certain Diptera, with reference to the heterochromosomes and the phenomenon of synapsis. J. exp. Zool. **5**, 359–374 (1908).

STOCKINGER, L.: Das Kernkörperchen. Protoplasma (Wien) **42**, 365–413 (1953).

STRASBURGER, E.: Zellbildung und Zellteilung. Jena: Fischer 1880.

STRASBURGER, E.: Über den Theilungsvorgang der Zellkerne und das Verhältnis der Kerntheilung zur Zelltheilung. Arch. mikr. Anat. **21**, 476–590 (1882).

STUBBLEFIELD, E.: The structure of mammalian chromosomes. Internat. Rev. Cytol., vol. 35 (G.H. BOURNE, J.F. DANIELLI, eds.), p. 1–60. New York: Academic Press 1973.

STUBBLEFIELD, E., WRAY, W.: Architecture of the Chinese hamster metaphase chromosome. Chromosoma (Berl.) **32**, 262–294 (1970).

SUMMIT, R.L., MARTENS, P.R., WILROY, R.S. JR.: X-autosome translocation in normal mother and effectively 21-monosomic daughter. J. Pediat. **84**, 539–546 (1974).

SUMNER, A.T., EVANS, H.J., BUCKLAND, R.A.: New technique for distinguishing between human chromosomes. Nature (Lond.) New Biol. **232**, 31 (1971).

SUMNER, A.T., EVANS, H.J.: Mechanisms involved in the banding of chromosomes with quinacrine and Giemsa. Exp. Cell Res. **81**, 223–236 (1973).

SUMNER, A.T., ROBINSON, J.A., EVANS, H.J.: Distinguishing between X, Y and YY bearing human spermatozoa by fluorescence and DNA content. Nature (Lond.) New Biol. **229**, 231–233 (1971).

SUN, N.C., CHU, E.H.Y., CHANG, C.C.: Staining method for the banding patterns of human mitotic chromosomes. Mammalian Chromosome Newsletter **14**, 26–28 (1973).

SWANSON, C.P.: Cytology and Cytogenetics. London: Macmillan 1960.

SZYBALSKI, W.: Properties and application of halogenated deoxyribonucleic acids. In: The Molecular Basis of Neoplasia. XV. Annual Symposium on Fundamental Cancer Research. Austin: Univ. Texas Press 1961.

TAKAYAMA, S.: The double strandedness of chromatid in the metaphase chromosomes of cultured cells treated with hot saline solutions. Jap. J. Genet. **48**, 1–9 (1973).

TANAKA, K., IINO, A.: Demonstration of fibrous components in hepatic interphase nuclei by high resolution scanning electron microscopy. Exp. Cell Res. **81**, 40–46 (1973).

TAYLOR, J.H.: Asynchronuous duplication of chromosomes in cultured cells of Chinese hamster. J. biophys. biochem. Cytol. **7**, 455–464 (1960).

TAYLOR, J.H.: The replication and organization of DNA in chromosomes. In: Molecular Genetics, vol. 1 (J.H. TAYLOR, ed.), p. 65–112. London-New York: Academic Press 1963.

TAYLOR, J.H.: The duplication of chromosomes. In: Probleme der biologischen Reduplikation. (Hrsg. P. Sitte), S. 9–28. Berlin-Heidelberg-New York: Springer 1966.

TAYLOR, J.H., WOODS, P.S., HUGHES, W.L.: The organization and duplication of chromosomes as revealed by autoradiographic studies using tritium-labeled thymidine. Proc. nat. Acad. Sci. (Wash.) **43**, 122–128 (1957).

TELLYESNICZKY, K. V.: Ruhekern und Mitose. Arch. mikr. Anat. **66**, 367–433 (1905).

TEPLITZ, R.L., GUSTAFSON, P.E., PELLET, O.L.: Chromosomal distribution in interspecific in vitro hybrid cells. Exp. Cell Res. **52**, 379–391 (1968).

THERMAN, E., SARTO, G.E., PATAU, K.: Center for Barr body condensation on the proximal part of the human Xq: A hypothesis. Chromosoma (Berl.) **44**, 361–366 (1974).

THERKELSEN, A.J., PETERSEN, G.B.: Variation in glucose-6-phosphate dehydrogenase in relation to the growth phase and frequency of sex chromatin positive cells in cultures of fibroblasts from normal human females and a 48-XXXY male. Exp. Cell Res. **48**, 681–684 (1967).

THOMAS, C.A., HAMKALO, B.A., MISRA, D.N., LEE, C.S.: Cyclization of eucaryotic deoxyribonucleic acid fragments. J. molec. Biol. **51**, 621–632 (1970).

THOMAS, C.A., ZIMM, B.H., DANCIS, B.M.: Ring theory. J. molec. Biol. **77**, 85–99 (1973).

THORLEY, J., WARBURTON, D., MILLER, O.J.: Absence of somatic pairing of sex chromatin masses (inactivated X-chromosomes) in cultured cells from a human XXXXY male. Exp. Cell Res. **47**, 663–665 (1967).

TJIO, J.H., LEVAN, A.: The chromosome number of man. Hereditas (Lund) **42**, 1–6 (1956).

TOLKSDORF, M.: The diagnosis of X-chromatin by the leukocyte test. In: Methods in Human Cytogenetics (H.G. SCHWARZACHER, U. WOLF, eds.). Berlin-Heidelberg-New York: Springer 1974.

TROSKO, J.E., WOLFE, S.L.: Strandedness of *Vivia faba* chromosomes as revealed by enzyme digestion studies. J. Cell Biol. **26**, 125–135 (1965).

TURPIN, R., LEJEUNE, J.: Les chromosomes humains. Caryotype normal et variations pathologiques. Paris: Gauthier-Villars 1965.

UNAKUL, W., HSU, T.C., RAO, P.N., JOHNSON, R.T.: Giemsa banding in prematurely condensed chromosomes obtained by cell fusion. Nature (Lond.) New Biol. **242**, 106–107 (1973).

UNNERUS, V., FELLMANN, J., CHAPELLE, A. DE LA: The length of the human Y chromosome. Cytogenetics **6**, 213–227 (1967).

UTAKOJI, T.: Differential staining patterns of human chromosomes treated with potassium permanganate. Nature (Lond.) **239**, 168–170 (1972).

VENDRELY, R., VENDRELY, C.: The results of cytophometry in the study of the deoxyribonucleic
acid (DNA) content of the nucleus. In: Internat. Rev. Cytol., vol. 5 (G.H. BOURNE, J.F. DANIELLI,
eds.), p. 171–228. London-New York: Academic Press 1956.

VOGEL, W., FAUST, J., SCHMID, M., SIEBERS, J.W.: On the relevance of non-histone proteins to
the production of Giemsa banding patterns on chromosomes. Hum. Genet. **21**, 227–236 (1974).

VOSA, C.G.: Heterochromatin recognition with fluorochromes. Chromosoma (Berl.) **30**, 366–372
(1970).

WAGENAAR, E.B.: End-to-end chromosome attachments in mitotic interphase and their possible
significance to meiotic chromosome pairing. Chromosoma (Berl.) **26**, 410–426 (1969).

WALDEYER, W.: Über Karyokinese und ihre Beziehung zu den Befruchtungsvorgängen. Arch. mikr.
Anat. **32**, 1–122 (1888).

WALEN, K.H.: Spatial relationship in the replication of chromosomal DNA. Genetics **51**, 915–929
(1965).

WALKER, P.M.B.: How different are the DNAs from related animals? Nature (Lond.) **219**, 228–232
(1968).

WANG, H.C., FEDOROFF, S.: Banding in human chromosomes treated with trypsin. Nature (Lond.)
New Biol. **235**, 52–53 (1972).

WARING, M., BRITTEN, R.J.: Nucleotide sequence repetition: A rapidly reassociating fraction of
mouse DNA. Science **154**, 791–794 (1966).

WASSERMANN, F.: Wachstum und Vermehrung der lebendigen Masse. In: Handbuch der mikroskopischen Anatomie des Menschen (Hrsg. W. V. MÖLLENDORFF). Berlin: Springer 1929. Bd. 1, Teil
2, S. 1–799.

WEISBLUM, B.: Why centric regions of quinacrine treated mouse chromosomes show diminished
fluorescence. Nature (Lond.) **246**, 150–151 (1973).

WEISBLUM, B., HAENSSLER, E.: Fluorometric properties of the bibenzimidazole derivate Hoechst
33258, a fluorescent probe specific for AT concentration in chromosomal DNA. Chromosoma
(Berl.) **46**, 255–260 (1974).

WEISBLUM, B., HASETH, P.L. DE: Quinacrine, a chromosome stain specific for deoxyadenylate-deoxy-
thymidylate—rich regions in DNA. Proc. nat. Acad. Sci. (Wash.) **69**, 629–632 (1972).

WENRICH, D.H.: The spermatogenesis of *Phrynotettix magnus* with special reference to synapsis
and the individuality of the chromosomes. Bull. Mus. comp. Zoolog. Harv. **60**, 57–134 (1916).

WETTSTEIN, R., SOTELO, J.R.: Fine structure of meiotic chromosomes. The elementary components
of metaphase chromosomes of *Gryllus argentinus*. J. Ultrastruct. Res. **13**, 367–381 (1965).

WHITE, M.J.D.: Animal Cytology and Evolution. Cambridge: Cambridge Univ. Press 1945.

WHITEHOUSE, H.L.K.: Towards an Understanding of the Mechanism of Heredity. 2nd ed. London:
Arnold 1969.

WICKBOM, T.: The time factor of chromosome spiralization. Hereditas (Lund) **35**, 245–248 (1949).

WILSON, E.B.: The Cell in Development and Heredity. 3rd ed. London: Macmillan 1928.

WOLF, U., FLINSBACH, G., BÖHM, R., OHNO, S.: DNS-Reduplikationsmuster bei den Riesen-Ge-
schlechtschromosomen von *Microtus agrestis*. Chromosoma (Berl.) **16**, 609–617 (1965).

WOLFE, S.L.: The fine structure of isolated metaphase chromosomes. Exp. Cell Res. **37**, 45–53
(1965).

WOLFE, S.L.: The effect of prefixation on the diameter of chromosome fibres isolated by the Langmuir-
trough-critical point method. J. Cell Biol. **37**, 610–620 (1968).

WOLFE, S.L., GRIM, J.N.: The relationship of isolated chromosome fibers to the fibers of the embedded
nucleus. J. Ultrastruct. Res. **19**, 382–397 (1967).

WOLFF, S., PERRY, P.: Differential Giemsa staining of sister chromatids and the study of sister
chromatid exchanges without autoradiography. Chromosoma (Berl.) **48**, 341–353 (1974).

WOLSTENHOLME, D.R.: Electron microscopic identification of sexchromatin bodies of tissue culture
cells. Chromosoma (Berl.) **16**, 453–462 (1965).

WYANDT, H.E., HECHT, F.: Detection of the X-chromatin body in human fibroblasts by quinacrine
fluoromicroscopy. Lancet **1971 II**, 1379.

WYANDT, H.E., IORIO, R.J.: Human Y-chromatin. III. The nucleolus. Exp. Cell Res. **81**, 468–473
(1973).

YAMASAKI, N.: Differentielle Darstellung der Metaphasechromosomen von *Cypripedium debile* mit
Quinacrin- und Giemsa-Färbung. Chromosoma (Berl.) **41**, 403–412 (1973).

YUNIS, J.J. (Ed.): Human Chromosome Methodology. London-New York: Academic Press 1965.

YUNIS, J.J., ROLDAN, L., YASMINEH, W.G.: Staining of repetitive DNA in metaphase chromosomes. Nature (Lond.) New Biol. **231**, 532–533 (1971).

YUNIS, J.J., SANCHEZ, O.: G-Banding and chromosome structure. Chromosoma (Berl.) **44**, 15–23 (1973).

YUNIS, J.J., YASMINEH, W.G.: Satellite DNA in constitutive heterochromatin of the guinea pig. Science **168**, 263–265 (1970).

YUNIS, J.J., YASMINEH, W.G.: Heterochromatin, satellite DNA and cell function. Science **174**, 1200–1209 (1971).

ZAKHAROV, A.F., BARANOVSKAYA, L.I., DEMINTSEVA, V.S., IBRAIMOV, A.I.: Differentiation along human chromosomes in relation to their identification. I. General pattern of 5-bromodeoxyuridine-induced differentiation. Tsitologiya **15**, 508–518 (1973).

ZAKHAROV, A.F., BARANOVSKAJA, L.I., IBRAIMOV, A.I., BENJUCH, V.A., DEMINTSEVA, V.S., OBLA-PENKO, N.G.: Differential spiralization along mammalian mitotic chromosomes. II. 5-bromodeoxyuridine and 5-bromodeoxycytidine-revealed differentiation in human chromosomes. Chromosoma (Berl.) **44**, 343–359 (1974).

ZAKHAROV, A.F., EGOLINA, N.A.: Differential spiralization along mammalian mitotic chromosomes. I. BUdR-revealed differentiation in Chinese Hamster chromosomes. Chromosoma (Berl.) **38**, 341–365 (1972).

ZAKHAROV, A.F., SELEZNEV, J.V., BENJUSH, V.A., BARANOVSKAYA, L.I., DEMINTSEVA, V.S.: Differentiation along human chromosomes in relation to chromosome identification. Excerpta med. (Amst.) **233**, 193 (1971).

ZANG, K.D., BACK, E.: Quantitative studies on the arrangement of the human metaphase chromosomes. I. Individual features in the association pattern of the acrocentric chromosomes of normal males and females. Cytogenetics **7**, 455–470 (1968).

ZANKL, Z., ZANG, K.D.: Structural variability of the normal human karyotype. Hum. Genet. **13**, 160–162 (1971).

ZECH, L.: Investigation of metaphase chromosomes with DNA-binding fluorochromes. Exp. Cell Res. **58**, 463 (1969).

ZEIGER, K.: Zum Problem der vitalen Struktur des Zellkernes. Z. Zellforsch. **22**, 607–632 (1933).

ZEIGER, K.: Physikochemische Grundlage der Histologischen Methodik. Dresden und Leipzig (1938).

ZENZES. M.T., WOLF, U.: Paarungsverhalten der Geschlechtschromosomen in der männlichen Meiose von *Microtus agrestis*. Chromosoma (Berl.) **33**, 41–47 (1971).

ZWEIDLER, A.: Die Replikation der Chromosomen von *Allium cepa*. Arch. Klaus-Stift. Vererb.-Forsch. **39**, 54–64 (1964).

Author Index

Page numbers in *italics* refer to the reference

Subject Index

Handbuch der mikroskopischen Anatomie des Menschen

7 Bände und Ergänzungsbände. Begründet von
W. VON MÖLLENDORFF. Fortgeführt von W. BARGMANN.

Band 1 Die lebendige Masse

Teil 1
Allgemeine mikroskopische Anatomie
und Organisation der lebendigen
Masse.
Vergriffen

Teil 2
Wachstum und Vermehrung der
lebendigen Masse.
1929. DM 290,—; US $ 118.90

Teil 3
Chromosomes in Mitosis and Interphase.
1976. DM 136,—; US $ 55.80

Band 2 Die Gewebe

Teil 1
Epithel- und Drüsengewebe. Binde-
gewebe und blutbildende Gewebe. Blut.
Vergriffen

Teil 2
Stützgewebe. Knochengewebe. Skelet-
system.
1930. DM 340,—; US $ 139.40

Teil 3
Gewebe und Systeme der Muskulatur.
1931. DM 140,—; US $ 57.40

Teil 4
Gewebe und Systeme der Muskulatur.
(Ergänzung zu Band 2/3)
1956. DM 70,—; US $ 28.70
Einbanddecke: DM 19,80; US $ 8.20

Band 3 Haut- und Sinnesorgane

Teil 1
Milchdrüse. Geruchsorgan.
Geschmacksorgan. Gehörorgan.
Vergriffen

Teil 2
Das Auge.
1936. DM 250,—; US $ 102.50

Teil 3
Die Haut. Die Milchdrüse.
(Ergänzung zu Band 3/1)
1957. DM 340,—; US $ 139.40

Teil 4
Das Auge und seine Hilfsorgane.
(Ergänzung zu Band 3/2)
1964. DM 440,—; US $ 180.40

Band 4 Nervensystem

Teil 1
Nervengewebe. Das periphere
Nervensystem. Das Zentralnerven-
system.
Vergriffen

Teil 2
Plexus und Meningen. Saccus vascu-
losus.
1955. DM 110,—; US $ 45.10
Einbanddecke: DM 19,80; US $ 8.20

Teil 3
Sensible Ganglien.
(Ergänzung zu Band 4/1)
1958. DM 320,—; US $ 131.20

Teil 4
Das Neuron. Die Nervenzelle. Die
Nervenfaser. (Ergänzung zu
Band 4/1)
1959. DM 490,—; US $ 200.90

Teil 5
Mikroskopische Anatomie des vegeta-
tiven Nervensystems. (Ergänzung
zu Band 4/1)
1957. DM 385,—; US $ 157.90

Teil 7
Hypothalamus. (Ergänzung
zu Band 4/1)
1962. DM 445,—; US $ 182.50

Teil 8
Das Kleinhirn. (Ergänzung
zu Band 4/1)
1958. DM 220,—; US $ 90.20

Teil 9
Allocortex. Bearbeitet von
H. Stephan.
1975. DM 680,—; US $ 278.80

**Band 5 Verdauungsapparat,
Atmungsapparat**

Teil 1
Mundhöhle. Speicheldrüsen. Tonsil-
len. Rachen. Speiseröhre. Serosa.
Vergriffen

Teil 2
Magen. Leber. Gallenwege.
1932. DM 230,—; US $ 94.30

Teil 3
Zähne. Darm. Atmungsapparat.
1936. DM 330,—; US $ 135.30

Teil 4
Die Leber-Gallengangsysteme,
Gallenblase und Galle.
(Ergänzung zu Band 5/2)
1969. DM 260,—; US $ 106.60

**Band 6 Blutgefäß- und Lymphgefäß-
apparat. Innersekretorische Drüsen**

Teil 1
Blutgefäße und Herz. Lymphgefäße
und lymphatische Organe, Milz.
1930. DM 230,—; US $ 94.30

Teil 2
Innersekretorische Drüsen I:
Schilddrüse, Epithelkörperchen.
Langerhanssche Inseln.
1939. DM 135,—; US $ 55.40

Teil 3.
Innersekretorische Drüsen II:
Hypophyse.
1940. DM 260,—; US $ 106.60

Teil 4
Innersekretorische Drüsen III:
Thymus, Paraganglien. Epiphyse.
Lymphgefäßapparat. (Ergänzung
zu Band 6/1)
1943. DM 250,—; US $ 102.50

Teil 5
Die Nebenniere. Neurosekretion.
1954. DM 540,—; US $ 221.40

Teil 6
Die Milz. Bearbeitet von F. TISCHEN-
DORF.
1969. DM 580,—; US $ 237.80

Band 7 Harn- und Geschlechtsapparat

Teil 1
Exkretionsapparat und weibliche
Genitalorgane.
1930. DM 250,—; US $ 102.50

Teil 2
Männliche Genitalorgane.
1930. DM 235,—; US $ 96.40

Teil 3
Weibliche Genitalorgane. Das Ova-
rium. (Ergänzung zu Band 7/1)
1957. DM 115,—; US $ 47.20
Einbanddecke: DM 19,80; US $ 8.20

Teil 4
Tube, Vagina und äußere weibliche
Genitalorgane. (Ergänzung zu
Band 7/1)
1966. DM 275,—; US $ 112.80

Preisänderungen vorbehalten/Prices
are subject to change without notice

Springer-Verlag
Berlin
Heidelberg
New York